FRAGONARD
MUSEUM
THE *ÉCORCHÉS*

CHRISTOPHE DEGUEURCE

FRAGONARD MUSEUM

THE *ÉCORCHÉS*

The Anatomical Masterworks
of Honoré Fragonard

WITH AN ESSAY BY
LAURE CADOT

Blast Books
NEW YORK

This book is dedicated to the memory of Pierre-Louis Verly, whose thesis for Doctor of Veterinary Medicine (1963), "Honoré Fragonard, Anatomist, First Director of the School of Alfort," was our vade mecum.

HONORÉ FRAGONARD ET SES ÉCORCHÉS © 2010 Réunion des musées nationaux, Paris

Translated from the French by Philip Adds © 2011 Blast Books, Inc.

English-language edition edited and designed by Laura Lindgren
French-language edition edited by Marie Lionnard with Cloé Brisset
Production manager: Philippe Gournay

Blast Books gratefully acknowledges the generous help of Eva Åhrén and Donald Kennison.

Pages 3 and 6: Honoré Fragonard, *The Horseman*, 1766–1771. Myology of a horse mounted by the myology of a man. No. 1322 in the 1794 inventory. ENVA, Fragonard Museum

Library of Congress Cataloging-in-Publication Data
Degueurce, Christophe, 1966–
  [Honoré Fragonard et ses écorchés. English]
  Fragonard Museum : the écorchés : the anatomical masterworks of Honoré Fragonard / Christophe Degueurce ; with an essay by Laure Cadot ; [translated from the French by Philip Adds]. — 1st ed.
    p. cm.
  Includes bibliographical references.
  ISBN 978-0-922233-39-7 (alk. paper)
1. Fragonard, Honoré, 1732–1799. 2. Anatomists—France—Biography. 3. Anatomical specimens—Catalogs and collections—France—Maisons-Alfort. I. Musée Fragonard (Maisons-Alfort, France) II. Title.
  QM16.F73D44 2011
  611.0092—dc22
  [B]   2010037235

Published by Blast Books, Inc.
P. O. Box 51, Cooper Station
New York, NY 10276-0051
www.blastbooks.com

Printed in France
First Edition 2011

10 9 8 7 6 5 4 3 2 1

Contents

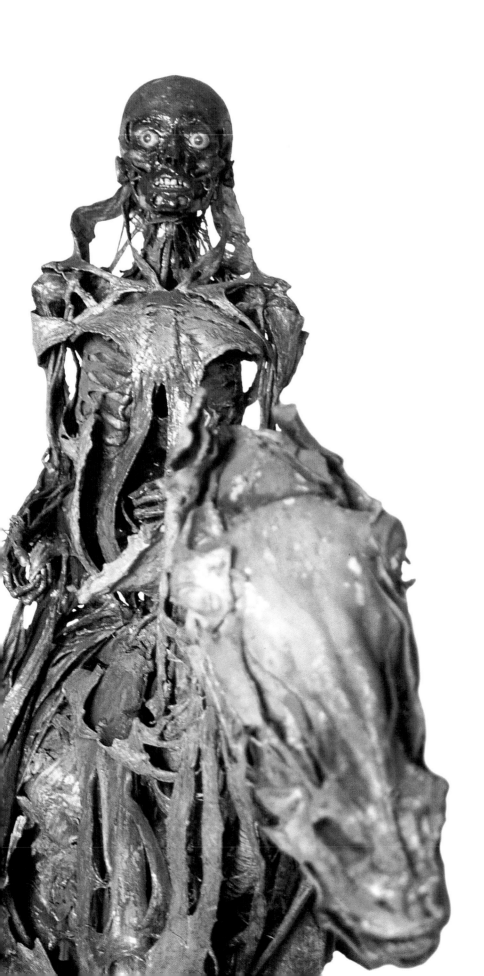

An Anatomist in the Century of Enlightenment

Surgeon, anatomist, and virtuosic preparator, the first practitioner of the art of dissection in the veterinary schools of Lyon and Alfort, Honoré Fragonard is known for his famous *écorchés*: twenty-one masterfully dissected human and animal specimens preserved more than two centuries ago and miraculously yet extant, very nearly all that remains of the horde of bodies dissected by his own hand.

The name *Fragonard* more likely brings to mind scenes of gallantry and feminine pulchritude, fair-skinned beauties amid lustrous silky fabrics, rather than parchmentlike dehydrated corpses. But the subject of this book is not Jean-Honoré Fragonard, the rococo painter who has given the name worldwide fame, but his cousin Honoré Fragonard, the anatomist. Because their names are so similar there is often confusion about the two cousins, both born in 1732 and both members of a prominent arts commission during the Revolution. Fragonard the anatomist has been overshadowed by his ever popular cousin, the painter of luminous palette whose work was the incarnation of the grace and the liberty associated with life in the eighteenth century. While one enjoyed a life of exuberant artistic creation and renown, the other was "continually hunched over cadavers," engaged in repugnant work in an icy dissection hall. To one, success and recognition for all posterity, to the other an unglamorous fate. It would not be until the nineteenth century that Fragonard the obscure anatomist-scholar of the Age of Enlightenment would emerge from the shadows, for until then the *écorchés* that had brought him acclaim in his lifetime lay forgotten in the collection of a private museum at the École nationale vétérinaire d'Alfort (National Veterinary School of Alfort). Human and animal cadavers strikingly posed; a flayed horseman riding into eternity mounted on a galloping horse; a dissected warrior with a menacing glare, brandishing the weapon of the biblical hero Samson; desiccated foetuses dancing a grotesque jig—these spectacular specimens, known today worldwide,

See p. 151

| 7

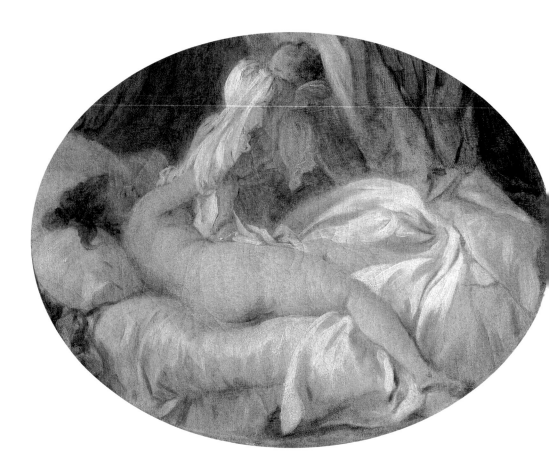

were as of only a few decades ago but obsolete relics that scientists ignored without, however, being able to bring themselves to destroy them. These bodies, intended to educate, have survived revolutionary upheavals, punishing environmental conditions, and indifference and remain today mostly intact, the prime examples of the art of dry preservation of anatomical specimens that were the pride of the "cabinets of curiosities" in the eighteenth century. The two Fragonards are symbols of the intellectual ferment that defined the Age of Enlightenment; Janus-like they face two directions, toward the beauty and joys of the world and toward an austere passion for science.

Fragonard the anatomist's life ended in 1799, at the close of a century that transformed the way people lived and thought. Political and social institutions as sources of knowledge had been turned upside down in the Enlightenment. Fragonard was part of a world hungry for learning and for liberation from ancient beliefs and simplistic dogma, a world that had seen the birth of a modern Homo sapiens. Organized by royal decrees, groups of intellectuals were brought together to accomplish the transformation. In a spirit of interdisciplinary study that prefigured modern science, the creation of the Académie des sciences in 1666 had given France a laboratory of thought, a meeting ground for the exchange of ideas that aroused aspirations for mankind. Louis XIV himself participated in the debate over the concept of circulation of the blood as published by William Harvey in 1628, and the king nominated his surgeon Pierre Dionis to the Jardin du Roi in 1672 to teach, against the tenets of the Faculté de médecine de Paris, "l'anatomie selon la circulation du sang" ("anatomy according to the circulation of the blood"). Fragonard was one of the craftsmen whose efforts accelerated the progress that had begun in the seventeenth century, and his works reflect the strife and challenges of his time.

In the medical world, the Age of Enlightenment was an extraordinarily fertile period of groundbreaking achievements. The founding of the Académie royale de Chirurgie (Royal Academy of Surgery) in 1731 had increased research activities and the exchange of information tenfold, and an unprecedented phase of intense scientific investigations ensued. Freed from an archaic teaching system, the way of disseminating increasing medical knowledge too evolved. The *écorchés* themselves attest to this transformation: these didactic works were created to demonstrate with utmost clarity to students the fundamental anatomical understanding to be gained from the cadaver. They embody the realization of an ideological shift begun by Vesalius in the sixteenth century: the triumph of direct observation of the dead and the living body alike over dogma and scholastic theory. A challenge to the dexterity of anatomists, the cadaver and anatomical specimens became the reference works for teaching at the same time that Latin had been superseded by modern languages. Despite the need for translations from one country to another, this change of language provided flexibility in the exchange of observations and descriptions, and it furthered a social revolution that opened the door of the medical and surgical professions to people of more modest backgrounds. Fragonard himself was a son of the people and his path was one that his ancestors probably could never have known. The new teachers in medicine, particularly in surgery, were often skilled, experienced practitioners; the scholars of philosophy who had occupied the faculties several decades earlier were now replaced by clinicians. The world of medicine and surgery was changing.

Ways of learning about the functioning of the body changed as well, and the eighteenth century saw the application of a logical method of observation followed by experimentation. Established facts led to further discoveries, and a new way of analyzing the mysteries of physiology and pathology arose. Molière's pedantic Dr. Diafoirus disappeared from the clinical schools as they reached their peak in the nineteenth century. The theory of humorism,

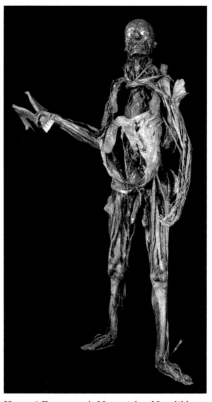

Honoré Fragonard, *Man with a Mandible*, 1766–1771, "myology and angiology of a man with a jawbone in his hand," no. 1149 in the 1794 inventory. ENVA, Fragonard Museum

Detail from
Man with a Mandible

the ancient medical notion correlating the organization of the human body (the microcosm—blood, phlegm, yellow bile, and black bile) with the functioning of the cosmos (the macrocosm—air, water, fire, and earth), gradually receded with the "invention" of the new realities of physiology.

The discovery in the eighteenth century of the laws of nature conflicted with divine law and gave rise to a school of thought that erupted several decades later in Darwinian theory. Suddenly, the natural world was no longer seen as a testimony to the grand scheme of a divine genius, the whole and perfect creation of an intelligent force. Nature acquired a coherence of its own independent of any divine intervention when the extensive commonality of the structure of living creatures was revealed by scholars such as Buffon, Daubenton, and John Hunter, who had organized his superb collection of comparative anatomy into an extraordinary functional continuum by arranging the organs function by function, from the plants up to man.[1]

Man and animal inexorably came closer together as this new understanding dispensed with old ideas that placed man above all other creatures; scientific revolution pulled him down toward the lower beings. In 1753, in *"Discours sur la nature des animaux"* ("Dissertation on the Nature of Animals"), included in the introduction to volume 4 of his *Histoire naturelle, générale et particulière avec la description du cabinet du Roi* (Natural History, General and Particular, with a Description of the Cabinet du Roi), Buffon suggested that natural history be applied in developing the noblest science, that of man.

Comparison of the anatomy of man and horse. François de Garsault, *Le nouveau parfait maréchal ou la connaissance générale et universelle* (A New Perfect Horseman, or The General and Universal Study of the Horse). Paris: Barois, 1755. ENVA library

SACKSICK

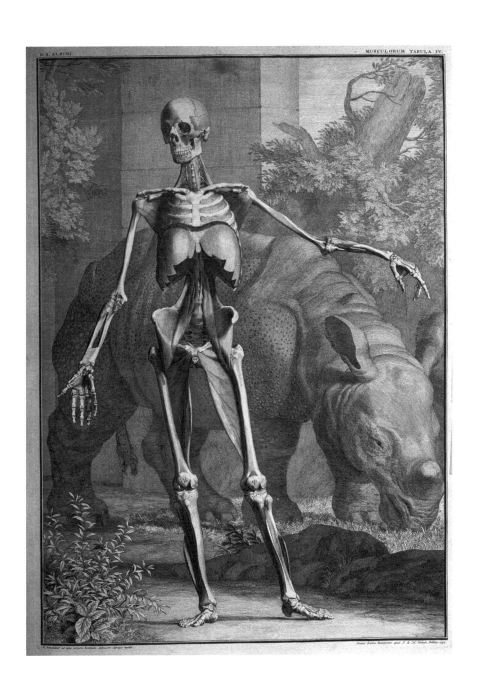

Si tabulas al quo corpore humano delineavit delineari curavit.

Prostat Lugduni Batavorum apud J. A. N. Verbeek, Bibliop. 1747.

Let us examine the nature of the animal world; let us investigate their organization, and study their general economy. This inquiry will enable us to draw particular inferences, to discover relations, to reconcile apparent differences, and, from a combination of facts, to distinguish the principal effects of the living machine, and lead us to that important science, of which man is the ultimate object.[2]

In the *Encyclopédie*, Claude Bourgelat, the future founder of the first two veterinary schools in the world, also pointed out in 1756 that progress in medicine, human and animal alike, go hand in hand.

These reflections do not detract from the analogy of the mechanism of the bodies of Man and the animal, which is truly constant. In traversing from the course of healing on the one hand and looking for new methods of treatment on the other, this fact is liable to get lost. Human medical science presents to the equine veterinarian abundant discoveries and riches, which must be put to good use. The physician must not flatter himself that he has mastered all, and the equine veterinarian when adequately skilled can in turn become a real expert.[3]

As an anatomist of man and animal, Fragonard found himself at the very center of this ideological whirlwind. Trained in the surgery of man, he started with human morphology as a reference point for the study of the ruminants, so familiar and yet paradoxically very little studied. He was one of the first to make the decisive step toward connecting human anatomy to the as yet uncharted field of animal systems. That one of his most impressive works—still extant—is of a man astride a horse is no mere chance. Displayed in this dissection is the musculature of these two species so closely linked in eighteenth-century economic, military, and social life but so dissimilar in morphology that Buffon juxtaposed them in his *Natural History* as two radically different creatures that work together. In the eighteenth century human and veterinary surgery became intertwined as comparative studies began to flourish and went on to inform all the natural sciences in the next century, which brought scientific, medical, and technical revolutions in France resulting from the work of men such as Claude Bernard, Pierre Rayer, and Louis Pasteur, whose discoveries changed the course of history.[4]

The *écorchés* are such finely crafted works that they have withstood—for more than two centuries—the ravages of time. Survivors of a world long gone, these last testimonies to a skill perfected in the Enlightenment and later abandoned can be appreciated both as remarkable permanent dissections created in the pursuit of practical anatomy and for an aesthetic power perhaps, or perhaps not, intended by their creator. Physical encounters with cadavers have been relatively rare in modern society, and seeing *écorchés* we inevitably experience a shocking reminder of the certain future that lies ahead of every person's life. Fragonard's heirs, anatomists of the twenty-first century silently at their work in laboratories—or, like Gunther von Hagens, exhibiting dissected bodies in popular exhibitions for the whole world to see—engage our desire to see or know, our repulsion, our reservations, just as does the work of their predecessors. Plastinates, the modern equivalent of the *écorchés*, today have been presented worldwide before an enormous public no less fascinated by the mysteries of the human body than in bygone times.

If placing Fragonard at the center of the history of science implies that *no* is the answer to the question so often asked by visitors to the Fragonard Museum—did Honoré Fragonard intend his *écorchés* as works of art?—we must yet reserve judgment. Since the sixteenth century, anatomists were dazzled by the delicate, complex machinery of the body, which they tirelessly explored in dissection. Likewise, the artists, draftsmen, engravers, and modelers after them considered the human body of first rank among the many marvels of an infinite nature. Scientific jubilation and aesthetic pleasure in praise of the divine creator can take many forms as, for example, in the well-known engraving by Albinus, professor of anatomy at the University of Leiden, in which the silhouette of a skeleton seems to lead a massive, docile grazing rhinoceros depicted in shades of gray. The artful composition and its underlying philosophical concept sacrifice none of its accuracy as a scientific representation of man and animal; rather, they enhance it.

Even if in our century science no longer considers the *écorchés* to be more than rather disturbing remnants of the long march toward human knowledge, the tracks of which must be carefully preserved, artists see in them dynamic and fertile aesthetic and conceptual associations. The bold posture of the *Man with a Mandible* has inspired the artist Gilles Sacksick to paint on a translucent canvas the tall stature of that imposing personage in broad black strokes. If intended by its creator as nothing more than a model demonstrating anatomy—a "beautiful specimen," as Fragonard noted—*Man with a Mandible* has validity as an artistic creation in its success as an anatomical model, in its refined, virtuosic execution, and in the powerful emotion it expresses. In the expansive dialogue between art and science, Fragonard certainly had much to say.

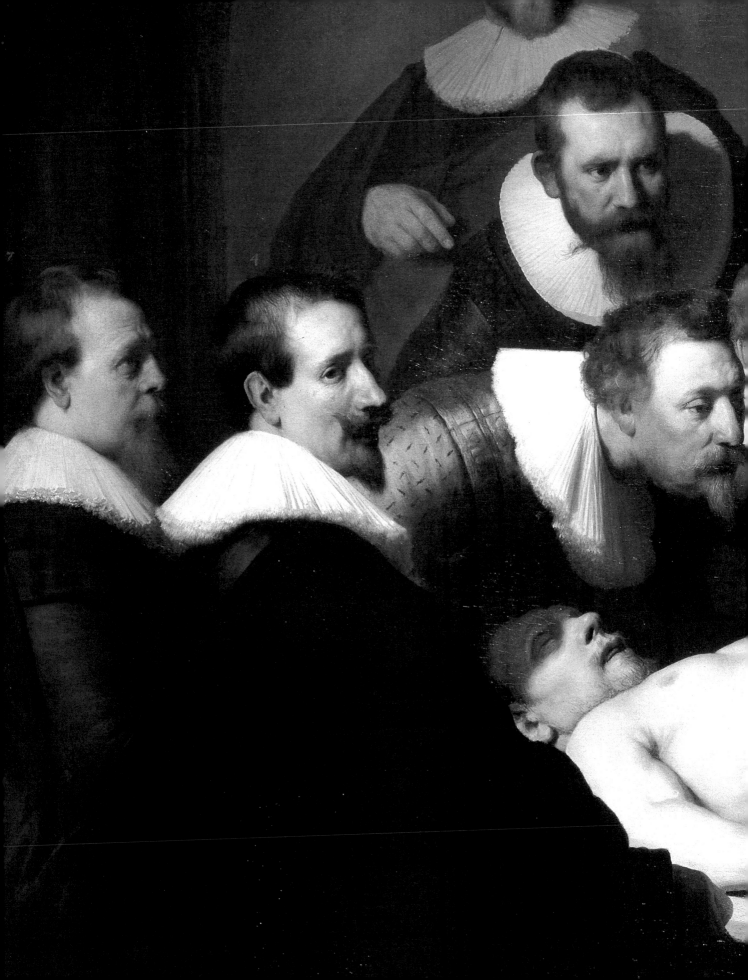

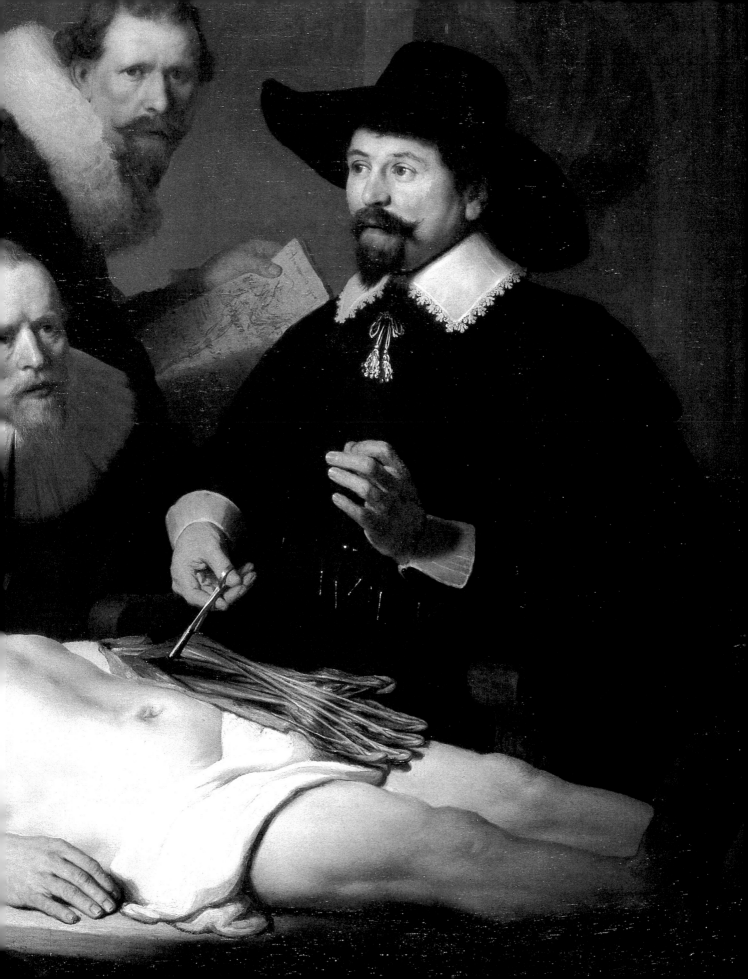

Honoré Fragonard
(1732–1799)

"Continually hunched over cadavers"

A SURGEON, SON OF A GLOVER AND PERFUMIER

Honoré Fragonard was born in Grasse, France, on June 13, 1732, into a family of Italian origin that had come to reside in Provence at the end of the sixteenth century.[1] His grandfather, treasurer of the Commune of Grasse and an oil merchant, was married in 1681 to Élisabeth Ricorde and they had seven children: first three daughters, then four sons, including Honoré, born February 1, 1694, the future father of the subject of this book. In a bourgeois house situated within the town precinct, Fragonard owned a shop where he sold gloves and perfume, and in the suburbs he also owned a large garden planted with jasmine and oranges, as well as a small house adjoining the garden in which he distilled the essence of seasonal flowers.

He married Marie Honorarde Isnard, daughter of master apothecary François Isnard, on November 8, 1723. They were to have five chil-dren: two daughters, Marie Élisabeth and Marie Anne, and three sons, Christophe Sauveur, Honoré, who was baptized on June 14, 1732, and lastly François, who later as a young man would follow in the footsteps of his elder brother Honoré in Alfort. Honoré Fragonard *fils* had a paternal uncle named François Fragonard, who married Françoise Petit in 1731, and their son the artist Jean-Honoré Fragonard (1732–1806) was born the following year on April 5 in Grasse, thus the two cousins were born just two months apart.

Their family tradition of showing their devotion to Saint Honorat (of whom the town cathedral preserved some relics) resulted in the two boys bearing practically the same Christian name. During the Revolution they were reunited in 1793 through the Commission Temporaire des Arts (temporary commission of the arts), and both were appointed as members of a jury panel headed by the painter

Honoré Fragonard, *Human Arm*, 1766–1771. ENVA, Fragonard Museum

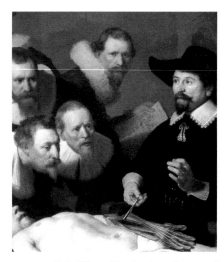

Detail from *The Anatomy Lesson of Dr. Nicolaes Tulp* (pages 16–17). Commissioned from Rembrandt by the powerful Guild of Surgeons of Amsterdam, this painting depicts an anatomy lesson in 1632 at the house of Dr. Tulp, a learned and eminent surgeon. The gaze of the spectators, fixed on the pale cadaver that the illustrious anatomist is dissecting, shows the intense interest aroused by this science.

Jacques-Louis David that conferred national prizes in painting, architecture, and sculpture.

Through his father's work, as a youth in Provence Honoré became familiar with botanical essences and their cosmetic application, learning techniques that doubtlessly later came to bear in his anatomical preparations.

Honoré Fragonard *père* died at age fifty-three on May 12, 1747, leaving Marie with great financial difficulties. Honoré had begun training as a surgeon, and he left home at eighteen for Lyon to further his studies. Two years later, his brother François, himself an apprentice surgeon, followed him to Lyon, but no trace of the two brothers is to be found in existing records until Honoré returned to his hometown in 1756 to begin an apprenticeship with the master surgeon René Lambert. A notarized certificate shows that in 1759, at age twenty-seven, he earned admission into the Grasse Guild of Surgeons.[2]

Fragonard was part of a new generation of practitioners who profited from an improved status in the medical world that had been hard won. The barber-surgeon, formerly subordinate to the physician (a learned Latin scholar immersed in the interpretation of the writings of the ancients), had become progressively more recognized by the authorities. In the Faculties of Medicine in the sixteenth century, a surgeon worked as a mere prosector under the direction of a professor who read from classical works. By the eighteenth century all that had changed. The Age of Enlightenment brought definitive advances, and out of the debris of an anatomical education both obsolete and ineffectual arose a surgeon who was a completely independent specialist. A series of royal decrees between 1723 and 1750 freed the master surgeon from the supervision of the physicians at public dissections and made theoretical and practical anatomical training mandatory. The foundation of the Royal Academy of Surgery, in December 1731, followed by the establishment in 1757 of the College of St. Côme, a school of practical dissection, completed the system for training newly independent surgeons.

If nothing appears to have predisposed Fragonard to study animal anatomy, he certainly had acquired extensive experience in the dissection and preparation of the human body by the time he encountered Claude Bourgelat (1712–1779). Bourgelat was the founder of the first veterinary school in the world, in Lyon in 1762, and although nothing is recorded of their first meeting it was pivotal to Fragonard's embrace of a new vocation and ultimate mastery of the art of the anatomical preparation of animals.

Claude Bourgelat was fifty years old when he and Fragonard met. He was a member of the gentry and a remarkable horseman who at age twenty-eight had become director of the prestigious riding academy of Lyon. His reputation in France and abroad had spread through the publication of his works on the equestrian arts and horse husbandry; between 1755 and 1757, he contributed 229 articles to the *Encyclopédie*, earning him the faithful friendship of its coeditor Jean D'Alembert (1717–1783). His aristocratic education and sharp intellect were equaled by his administrative skills, as he demonstrated over several years as director of the library and inspector of stud farms at Lyon. The study of diseases of the horse was still quite primitive, being in the hands of uneducated people who came to be called *empiriques*, not qualified trained veterinarians. A visionary pioneer in veterinary science, Bourgelat wrote, between 1750 and 1753, a seminal treatise in horse husbandry, the *Élemens d'Hippiatrique*, and in 1760 he created a school of farriery (horse shoeing) in Lyon, before his connections with the intellectuals of the day helped him to realize his dream of creating a veterinary school. Two political figures were instrumental in his successful establishment of a place for veterinary instruction that received protection of government authorities. Chrétien Guillaume de Lamoignon de Malesherbes (1721–1794), director of the library, friend of the encyclopedists, and Henri-Léonard Bertin (1720–1792), intendant (administrator) of Lyon and later contrôleur général des finances (controller-general of finance) for Louis XV, recommended the project to the king, citing economic gains in livestock of all breeds—bovine, ovine, and caprine.[3] The official opening of the veterinary school of Lyon on January 1, 1762, was the result of the work of Bourgelat, Malesherbes, and Bertin.

There were eighty-eight students enrolled in Lyon as of 1764 when the opening of the School of Alfort hindered its growth. Prospective students were required only to know how to read and write, and there was no minimum age limit. Fees were usually covered by the French province or country from which the students came; few were able to pay their own way. From the outset, a grant system supported the training of young men of humble background who would later return to their communities. The demonstrators were Pons and Fragonard; surgery was taught by Claude Flurant (1721–1779), farriery by Philibert Chabert (1737–1814), and botany as well as materia medica by the abbé Rozier (1734–1793), a botanist and agronomist later of great fame. Bourgelat did

Copy by P. Cohendy, 1907, of Vincent de Montpetit, *Portrait of Claude Bourgelat*, 1752–1776. Painted in 1752 in Lyon, the work was revised in 1776 in Paris to update the subject's face. ENVA

Alexandre Roslin, *Portrait of Henri-Léonard-Jean-Baptiste Bertin*, 1768. Château de Versailles

not himself teach but rather discharged the duties of director. He sent the
students into areas experiencing outbreaks of livestock epidemics and
publicized their successes, more or less proven, which served to promote
his school.

Why Fragonard went to work at the school after having already been
a surgeon for three years remains unknown. He is named in the foundation of the school as professor and demonstrator of anatomy, and account
ledgers of the veterinary school from May 1, 1763, list him as the director.
Already chair of anatomy, Fragonard was also made responsible for
settling the monthly accounts for the pharmacy, the stables, and the blacksmiths, for which he received a salary increase. In 1765 he earned 1,200
livres, Bourgelat 5,000.[4]

Aside from miscellaneous administrative reports, there exist very few documents between 1763 and 1765 relating to Fragonard. It appears that he held classes involving the assiduous study of cadavers, as later was done at Alfort, mainly in the practical room. A great number of specimens were preserved by the hand of Fragonard during this time. They served both in course demonstration and in the publication of Bourgelat's *Précis anatomique du corps du cheval* (Handbook of the Anatomy of the Horse) in 1768, which, it must be said, unscrupulously appropriated the work of Fragonard. Specimens were being preserved by desiccation just as they were later at Alfort and, be it myth or reality, various records relating to this first veterinary school mention "an immense preparation of a dissected man mounted on a horse with its superficial muscles likewise displayed,"[5] as reported in the 1782 catalogue of the Cabinet d'histoire naturelle at Lyon. It thus appears that Fragonard had already created an example of his famous dissected *Horseman*, on display today in the Fragonard Museum at Alfort. That an anatomical collection had been established at Lyon soon after the school opened is certain, since numerous specimens accompanied Fragonard's arrival at the new school at Alfort upon its foundation.

In 1764 the king bestowed on Bourgelat the title inspecteur général de l'école vétérinaire de Lyon et de toutes les écoles vétérinaires à établir dans le royaume (inspector general of the veterinary school of Lyon and of all the veterinary schools to be established in the realm), a title indicating an intention to expand such schools into all the provinces of France. Bourgelat soon undertook to found a school in Paris, where he was also conferred a new function of commissaire général of the royal studs. He left Lyon on April 15, 1765, and three months later Chabert and Fragonard joined him in Paris in a temporary location nearby Porte Saint-Denis. Fragonard resumed his work in the following winter. Found among a list of the cost of various utensils, drugs, and chemical substances, in a "statement of accounts made to the Pavillon Mazarin" on January 11, 1766, is, for example, "for a donkey sent to Monsieur Fragonard for dissection and for oil to be used in the injection, 4 livres." With his young students at the temporary site in Paris, Fragonard was again dissecting and creating anatomical specimens, and as he had done at Lyon he made several mounted specimens and collected all sorts of material valuable in the teaching of anatomy.[6]

Meanwhile Bourgelat negotiated the purchase of a property in Alfort that seemed ideal for the new school he had left Lyon to establish. Twenty-five acres (10.5 hectares) just outside the suburbs of Paris, the

Veüe de Halfort de deſſus le pont de Charanton

B. Flomen jn. et ſe.

Auec priuil. du Roy. 4

The château of Alfort c. 1670
Archives départementales
du Val-de-Marne, Créteil

Alfort property lay nearby the bridge at Charenton. It was sited in a farmland area stocked with cattle and horses, ideal for supplying study subjects. The property included the château, a large house fronted by a portal; several outbuildings; gardens; and a stretch of fallow grounds. Lodgings for the pupils as well as a hospital for eighty animals were needed by autumn 1766.[7] This accelerated building of the school was accomplished by the famed architect Jacques-Germain Soufflot (1713–1780), architect of the Panthéon in Paris and named by the king as contrôleur and later intendant général des bâtiment (director of the king's buildings).

Six students were enrolled as of the official opening of the school at Alfort in the first week of October 1766. Among them was Fragonard's brother François, who like Honoré had become a surgeon. As at Lyon, Bourgelat recruited sturdy young men accustomed to hard physical farmland labor and able to handle a horse. These new students, generally the sons of farmers and of blacksmiths, may have been capable in the art of the forge and in shoeing horses, fundamental in equine veterinary study, but their education was generally so minimal that they could scarcely read or write. Bourgelat hoped such recruits would later return to their home provinces to exercise their new skills, as was too the wish of the government authorities. Betraying his deep distrust, Bourgelat insisted that doctors and surgeons, and in particular the gentry, were not suited to work in the forge or hospital. Could it be that he felt a tinge of threat of competition from educated people who might challenge his authority or seek to open rival schools in other towns?

The staff at Alfort was small. Besides Bourgelat and Fragonard there were a pharmacist, a demonstrator of anatomy, a gardener, and, lastly, team leaders chosen from among the students as supervisors and teaching assistants. Chabert was soon named professor of farriery. As at Lyon, Bourgelat did not teach; his aristocratic background set him apart from such duties, but he devised the curriculum for each discipline and kept a close watch on the students' progress. He oversaw the publication of the programs of study, using the series of handbooks he had prepared at Lyon as the basis for a tightly structured education that in the long run was regarded as rigid and stifling.

A prerequisite to learning pathology, the study of anatomy aimed to impart to its pupils a complete understanding of the body of not only the horse, the ox, or the ewe, but also the human body, a point of reference in comparative anatomy that was much explored during the eighteenth century. The study of cadavers was hands on; to demonstrate delicate structures, Fragonard employed the injected and dehydrated anatomical preparations that he had continually prepared for expressly that purpose. Students who showed a particular aptitude for specimen preparation and taxidermy or mummification received special additional training in these techniques, and the anatomical specimens they produced were placed on display in what came to be known as the school's "cabinet." One particularly outstanding student, Jacques-Marie Hénon (1749–1809), the son of a poor farmer from Picardie, became Fragonard's principal collaborator. Hénon was replaced following Fragonard's eventful dismissal,

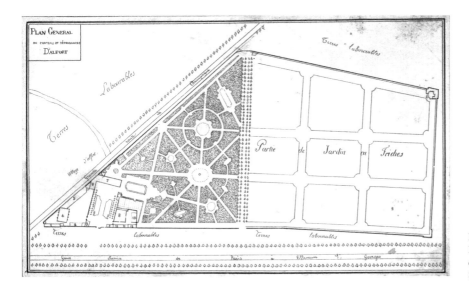

Plan of the Royal Veterinary School of Paris, 1766. A. Railliet and L. Moulé, *Histoire de l'École d'Alfort* (History of the Veterinary School of Alfort), 1908, p. 25, fig. 6

and in 1780 he too left Alfort for Lyon, where he taught until the end of his career.

On December 27, 1766, the King's Council established a chartered diploma in the art vétérinaire, creating the profession of veterinarian for students who successfully completed four years of study. The *empiriques* and the gelders were still free to carry on their activities, but only "chartered" practitioners were granted the title of veterinary surgeon. A ban on *empiriques* treating animals was not proclaimed until 1938, and it was not enforced until the 1970s.

Bourgelat governed over all aspects of life at the school and all the details of its organization. His rigid and fastidious *règlemens* of 1777 evidence his obsession with discipline.

The eminent staff and intellectual aura of Alfort and Lyon made them the models of veterinary science education throughout all of Europe. Many foreigners enrolled and on returning to their homelands participated in the founding of new schools, which in turn spawned yet more schools. After Alfort, schools were established first in Turin, in 1769, followed by Göttingen (1770), Copenhagen (1773), Vienna (1777), and the Royal Veterinary College in London was opened in 1791 by Charles Vial de Saint-Bel (1750–1793), friend of the celebrated John Hunter (1728–1793).

Main room of the Fragonard Museum, 2010

THE ESTABLISHMENT OF THE CABINET OF ALFORT BY FRAGONARD AND HIS PUPILS

As at Lyon, Claude Bourgelat created a cabinet at Alfort far surpassing any simple collection of natural history. It was a veritable museum housing the widest variety of specimens: intestinal parasites, monstrous foetuses, pathological osteology, and remarkable digestive stones, among others. Installed on the first floor of the building along the road to Champagne, now the avenue du Général-Leclerc, the cabinet received official royal status in the momentous proclamation of 1777, after Fragonard's departure: "The preparations which are to be made in the schools shall be displayed in a cabinet dedicated to the glory of his Majesty, to demonstrate the gratitude of the schools and the agriculturalists to him. It shall be named the Cabinet du Roi and its supervision is to be entrusted expressly to the professor of anatomy."[8]

From the outset the cabinet was open to the public. At a time when no such other large comparative anatomical collection was available to the citizens of Paris, at Alfort anatomists, doctors, and famous naturalists mingled with the curious visiting the collection. Its fame spread throughout Europe, and numerous references to it appear in travelogues and guidebooks listing the bountiful tourist attractions of the era.

The *Almanach vétérinaire*[9] of 1782 gives the most detailed description.

In large cabinets standing all around the antechamber are a quite large quantity of quadruped animals, of birds from the four corners of the world, stuffed and conserved with the greatest care; very precious injected specimens, stones of considerable size found in the intestines of horses, calculi, pebbles, etc.

The next room which follows is immensely long and is appointed with four large display cases placed at the corners, each 7½ feet long by 3½ feet wide [2.5 m x 1.16 m], enclosed with glass doors.

On the right and between the windows, facing the notice boards, are seven more small cases of the same design. On the left, facing the casements, stands a display cabinet occupying nearly the whole of the length of the room, filled, like the cases, with every example of the greatest skill imaginable in the art of preserving animals. Thirty perfectly complete pieces reveal to the viewer every single part of the animal; comparative specimens abound; the thinner parts are injected and preserved; a bovine stag, an ox, a llama, a vicuna, some François de Cange stags, bulls, Angolan rams, foxes, dogs, camels, tigers, panthers, eagles, peacocks, chickens, etc.—all are placed in opposition with the

most primarily useful of animals to mankind, the horse, which is itself opposed for comparison with the superb specimens of the human body. Still more cases are pleasingly arranged in the middle of the room. Atop these cabinets are skeletons of various animals, and in the center of the great display cabinet rests a bust in relief made of porcelain, imitating bronze, a very good likeness of King Louis XV, donated to the school by the Manufacture de Sèvre.

Above a door at back, intended to lead to a planned third room, is a crest painted in tempera in which two nymphs obscure the Royal coat of arms, decorated with a ribbon of His Orders. As one nymph scatters flowers across the shield, the other leans with her elbows on the shield as a support, gesturing to the first to place the flowers lower down; in her left hand she holds a crown of oak, symbol of the riches that await the Artistes Vétérinaires. Opposite this door, the main door too is decorated overhead with a painted crest. Shaded by two cock's wings and surmounted by a royal crown, an antique shield displays a royal crest formed by two interlaced branches of oak and two palms knotted and tied by a ribbon and attached to the shield, supported on either side by agricultural instruments. Pictured in the crest are some small children playing with a plough and the skull of an ox.

Another document, the *Almanach du voyageur à Paris*,[10] published in 1785 by Luc Vincent Thiéry (1734–1811), reported, "the royal veterinary school in the château at Alfort near Charenton possesses a superb anatomical collection, which has become the greatest of its kind in all of Europe."

Among the intellectuals who flocked to Alfort from all over Europe were the German surgeon Georg Ludwig Rumpelt, professor of veterinary medicine at Dresden and chief veterinary officer of the royal stables of the prince of Saxony. Visiting the cabinet at the end of 1779 on a fact-finding mission in France,[11] he left rare, treasured remarks on the role and work of Fragonard.

Rumpelt's compatriot Heinrich Sander, a twenty-five-year-old naturalist and professor at the Gymnasium illustre of Karlsruhe, methodically explored all the cabinets of curiosities of Paris and wrote a detailed description of each room at Alfort.[12] These eyewitness accounts became invaluable after Bourgelat eradicated all credit to the anatomist of Alfort—Fragonard—for his creations after his departure from the school.

Later, in 1802, after revolutionary turmoil robbed the cabinet of a large part of its collection to profit the École de santé (School of Health) in Paris, the zoologist Karl Asmund Rudolphi (1771–1832), a great name in the field of veterinary medicine from its inception, furnished a crucial description[13] of all the specimens remaining in the plundered cabinets at Alfort.

Main room at the Fragonard Museum, 2010. Jacques-Nicolas Brunot, *Bust of the Stallion "Truffle,"* 1828. Truffle was a Thoroughbred owned by King Charles X of France (1757–1836).

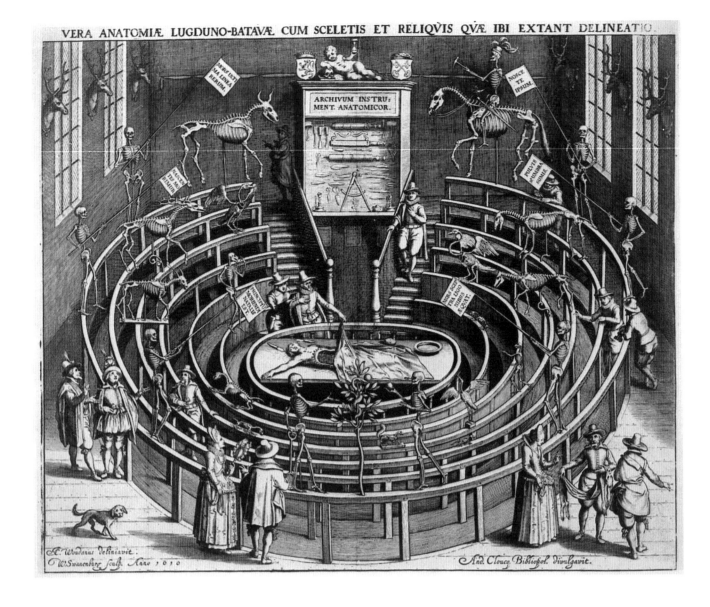

VERA ANATOMIÆ LUGDUNO-BATAVÆ CUM SCELETIS ET RELIQVIS QVÆ IBI EXTANT DELINEATIO.

THE CABINET AT ALFORT IN THE CONTEXT OF
ANATOMICAL AND NATURAL HISTORY COLLECTIONS

Wilhelm Swanenburg's 1610 engraving of the anatomical theater installed at Leiden, Netherlands, by Pieter Paaw (1574–1617). The dissection session, often open to the general public, was sanctified by its decor. The wooden benches surrounding the center, where the corpse is being dissected, are adorned by numerous animal and human skeletons, including a human skeleton mounted on that of a horse. Leiden Museum, Boerhaave

Fragonard was by no means the first European anatomist to create a collection combining anatomical and natural history specimens. The practice of assembling both documentary and pedagogical collections arose naturally as the science of anatomy, which conducted research directly on the cadaver, developed. Dissections took place in anatomical theaters furnished with interesting specimens relevant to the nascent science, and many textbooks embellished with woodcut illustrations appeared. As methods of preservation that had been developed in the seventeenth century were mastered, great anatomical collections arose.

Direct investigation into the human body seems to have begun in Bologna as early as 1270, in dissections undertaken in academic study.

The first known anatomical observation referencing a cadaver was made by the Italian physician Mondino de Luzzi (c. 1270–1326) in his *Anathomia* written around 1316, first printed in 1478,[14] and used for some three centuries. In this Galenic treatise, the first modern dissection manual, Mondino describes the dissection of two female cadavers. A woodcut depicts an enthroned professor reading aloud from a book as a prosector, probably a surgeon, carries out the dissection and an assistant uses a wand to point to the anatomical structures described in the scholarly text.

It was Andreas Vesalius (1514–1564) who revolutionized anatomy with his momentous *De humani corporis fabrica* published in 1543. This enormous work changed the study of anatomy, Vesalius having disproved Galen's and Mondino's work, and it prefigured the opening of many anatomical theaters in which a collection of objects such as skeletons or ornamental engravings would be displayed, as well as the dissected body.

Anatomical collections displayed in dedicated museums had to be well preserved, practical, and pleasing to the eye. Vascular injections of wax in contrasting colors greatly enhanced the appearance of specimens, making them quite beautiful. The Dutch naturalist Jan Swammerdam (1637–1680) appears to have been the first to succeed in making such injections in the 1660s. He published the method in his *Miraculum naturae sive uteri muliebris fabrica* of 1672, with a supplement presenting "a new method of preservation to permanently retain the form and appearance of hollow parts of the body." His collection became so famous that the grand duke of Tuscany offered 12,000 florins for its purchase—an offer that was rejected.[15]

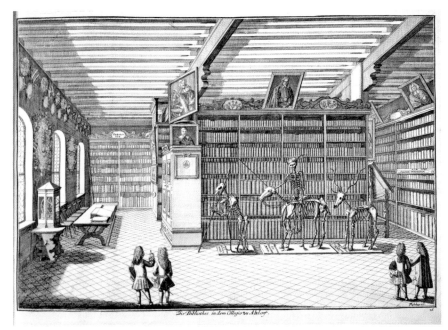

Johann G. Puschner, *Amoenitates Altdorfinae oder Eigentliche nach dem Leben gezeichnete Prospecten der Löblichen Nürnbergischen Universität Altdorf* (Amoenitates Altdorfinae, or Scenes of Life at the Laudable Nuremberg University at Altdorf Drawn from Life), Nuremberg, c. 1720, plate 16, the library of the University of Altdorf. The skeleton of a man mounted on the skeleton of a horse is flanked by the skeleton of a bear (left) and stag (right). The Altdorf rider's skeleton, clutching a spear, belonged to a Croatian terrorist from Nuremberg. With his *Horseman*, Fragonard re-created the classic anatomical figure of a man on horseback. Universitätsbibliothek, Erlangen

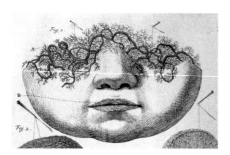

Frederik Ruysch, *Thesaurus anatomicus tertius*. Amsterdam, *apud* Joannem Wolters, 1703, plate 4, fig. 1. Engraving by Cornelius Huÿberts. Injection of the arteries of the head of an infant, the craniofacial bones removed. ENVA library

It was one of Swammerdam's contemporaries and colleagues who perfected the art of injection: Frederik Ruysch (1638–1731), the botanist and anatomist particularly acclaimed for his displays of human and animal remains. Posing foetal skeletons amid hardened injected arteries and urinary calculi, Ruysch composed macabre tableaux based on mythology, musical production scenes, and the like. He was especially well known for his fluid-preservation technique, suspending in his own secret solution not only body organs but babies and small children, the faces of whom were so lifelike it seemed they might awaken and speak. Beginning in 1689, Ruysch built up a vast collection, which he organized into ten cabinets, and he published descriptions of them in separate volumes. The greater part of his collection was acquired in 1717 by Peter the Great and today remains preserved in the Hermitage Museum.

. . .

Frederik Ruysch, *Thesaurus anatomicus primus*. Amsterdam, *apud* Joannem Wolters, 1701, plate 1. Engraving by Cornelius Huÿberts. Assemblage of foetal skeletons with various calculi, injected and hardened veins and arteries, and a handkerchief made from human peritoneum. ENVA library

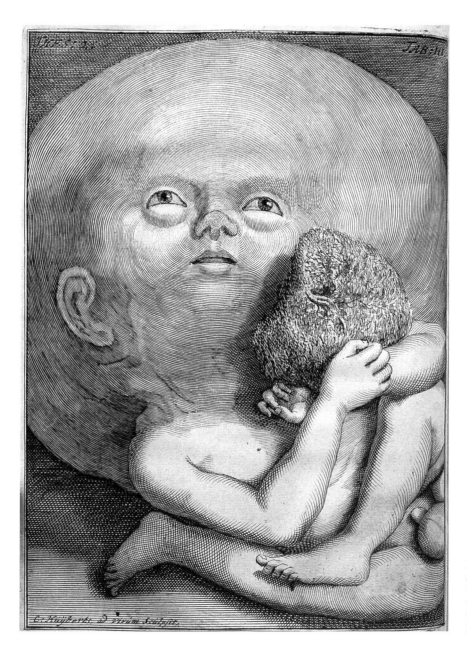

Frederik Ruysch, *Thesaurus anatomicus secondus*. Amsterdam, *apud* Joannem Wolters, 1702, plate 3. Engraving by Cornelius Huÿberts. Hydrocephalic infant holding part of its own placenta. ENVA library

Among the great preparators of history, the Hunter brothers, William and John, each of whom founded his own museum and became very well known in the eighteenth century, are mentioned by Fragonard himself in his letter of 1792 to the national assembly. The elder, William Hunter (1718–1783), left his native Scotland for London, where he established himself as an anatomist and leading obstetric consultant. In 1768 he created an important anatomy theater in London where he amassed a collection of minerals, numerous human specimens, and several specimens of comparative

FR. RUYSCHII OPERA OMNIA

AMSTELÆDAMI apud JANSSONIO-WAESBERGIOS 1720.

anatomy. He died at age sixty-four with no heir, leaving his collection to the University of Glasgow, where he had begun his career. William's brother John Hunter, ten years younger and described as a kind but brusque man of little culture, joined William in London in 1748. He became William's prosector and acquired so thorough a knowledge of anatomy that he was elected a master of anatomy at Surgeons' Hall in 1753. John Hunter was passionate about human and comparative anatomy, always seeking to understand the relationship between structure and function and to discover commonality of systems among all living creatures. He started the first museum of comparative anatomy, from plants to humans and insects to whales. A place of study and research, it was also open to the public two months a year. John's wish that his work should be preserved after his death, at age sixty-five, was realized by its purchase by the government in 1799 for £15,000, which placed it in the care of the Company (later the Royal College) of Surgeons in London.

This collection numbered around 15,000 specimens from more than five hundred species of animals and plants and contained many hundreds of spirit-preserved specimens. The Hunterian Museum's collection today contains approximately 3,500 of Hunter's original specimens and preparations among its many thousands of items, much of the original collection having been destroyed by wartime bombs in 1941.

France was not far behind England, with the Jardin du Roi and its important collection of specimens at the forefront of research. Founded in 1635, the Jardin royal des plantes médicinales (Royal Medicinal Plant Garden) was originally intended for physicians and apothecaries engaged in studying the identification and therapeutic use of plants. A 1718 proclamation by the boy-king Louis XV broadened its purpose to include natural history and chemistry, and it became known as simply the Jardin du Roi; in 1729 the name Cabinet d'histoire naturelle was also associated with the Jardin du Roi.

Georges-Louis Leclerc, Comte de Buffon (1707–1788), was appointed director in 1739 and held that position for almost fifty years, until his death. Buffon surrounded himself with the best naturalists, doubled the size of the garden, and significantly enlarged the collection, whose treasures, confined to two rooms, were increasingly crammed together. In 1749 the first volume of his *Natural History* was published. An encyclopedic work eventually comprised of forty-four volumes (thirty-six in his lifetime), it was the first modern attempt to present a systematic account of the whole of nature. Its last volume was published in 1804. Buffon completed the trove at the Jardin du Roi by acquiring major depositories such as the collection (still on display within the Jardin des plantes as the Cabinet

Portrait of John Hunter. Published in the atlas Œuvres complètes de John Hunter (Complete Works of John Hunter), by Gustave-Antoine Richelot. Paris: Fortin Masson et Cie, 1843. ENVA library

OPPOSITE
Frederik Ruysch, frontispiece to *Opera omnia anatomico-medico-chirurgica.* Amsterdam, *apud* Janssonio-Waesbergios, 1721. Engraving by Cornelius Huÿberts. In this depiction of Ruysch's cabinet, the allegorical figure Time pays homage to Mother Nature with a cornucopia, surrounded by dried and fluid-preserved specimens. ENVA library

Alexandre Roslin,
Portrait of Daubenton, 1793.
Musée des Beaux-Arts, Orléans

Bonnier de la Mosson) of the wealthy Joseph Bonnier de la Mosson (1702–1744), an aristocrat and great lover of art and the sciences, and the collection of the scientist and naturalist René-Antoine Ferchault de Réaumur (1683–1757), sometimes called the Pliny of the eighteenth century.

Buffon steadfastly directed the institution toward exhibition, yet he was criticized for neglecting less presentable specimens that were hidden away in storage and seldom used. In 1745 Buffon recruited a brilliant assistant, Louis-Jean-Marie Daubenton (1716–1800), who later became professor of rural economy at the Veterinary School of Alfort in its glory years of the 1780s. Daubenton's major contribution in the development of the collections was their systematic organization, and he collaborated on Buffon's *Natural History* and published several important works, among them a description of the Cabinet du Roi (1749–1767), one of the earliest significant works of museology. His particularly notable collection of skeletons of quadrupeds was unfortunately jammed into a tiny room and not seen by the public until the beginning of the nineteenth century. Were it not for the successful intervention of Daubenton and of the botanist and agronomist André Thouin (1747–1824), a friend of the U.S. president Thomas Jefferson, the Jardin du Roi could easily have disappeared during the Revolution as did many other royal institutions. By decree of the Convention dated June 10, 1793, the Jardin du Roi became the Muséum national d'histoire naturelle and was placed under Daubenton's direction. The new museum immediately acquired property from expatriates as well as part of the collection of the Cabinet of Alfort and donations from merchants in direct contact with distant lands.[16]

The richest anatomical treasures, however, were not always to be found in public institutions. Among the wealthy, both aristocratic and bourgeois, the possession of a collection of "curiosities" was considered a mark of cultivation, demonstrating one's interest in the spectacular new scientific advances and philosophical ideas of the eighteenth century and reflecting one's wealth. These *cabinets des curieux*, as Fragonard called them, included examples from every scientific discipline. Cabinets were formed around optics, astronomy, or chemistry, but most exhibited rarities plucked from the fascinatingly diverse natural world in grand explorations. Taxidermic animals, corals, minerals, and dried plants were displayed alongside dried, preserved actual human body parts or models, made perhaps of wax. Prices paid to satisfy a passion to collect items from this exciting new world were high. The royal family too was keen to collect: Louis XV installed cabinets at Versailles, and his cousin the duke of Orléans amassed what was probably the most beautiful collection anywhere of artificial anatomical material.

The rise of the Jardin du Roi and of the veterinary schools, and the vogue for cabinets of curiosities, are indications of the intellectual and institutional trends that inspired Bourgelat in 1766, then Fragonard in 1792, to work toward founding a Cabinet national d'anatomie (national anatomy museum).

FRAGONARD'S RENOWN AND DISMISSAL

By 1770 Fragonard began to gain a well-deserved renown for his works, and even Henri Bertin, minister of state under Louis XV and government overseer of the veterinary school, greatly admired his skill.

At Alfort, Fragonard was engaged in several fields of activity. Important discoveries in the therapeutic properties of plants in the eighteenth century led Bourgelat and his successors to give botany prominence in the curriculum. As at Lyon, Fragonard oversaw maintenance of the garden; he was in all likelihood at the same time engaged in demonstrations of materia medica. The pharmacy provided a good source of income for the new establishment, which profited from the public sale of medications. Fragonard was also in charge of the workshop in which his brother François worked.

Clashes between François and Bourgelat, however, ended finally in François's suspension in 1770. He left the school and established himself as a master surgeon in Pecq, near Saint-Germain-en-Laye.[17] This was but one episode in an ever increasing tension between the despotic founder of the school and its professor of anatomy. Like the other teachers at Alfort, Fragonard tolerated with difficulty Bourgelat's petty authoritarianism, and the disagreements between two such different men could only become worse. Fragonard likely nursed a certain bitterness toward his superior particularly since Bourgelat had, among other outrages, published *Précis anatomique du corps du cheval* in 1768 without so much as mentioning Fragonard's important contributions. In 1771 their mutual animosity became yet more pointed: the inspector general accused Fragonard of showing him too little respect, and, citing serious negligence in his roles as director and demonstrator, Bourgelat obtained Fragonard's dismissal in early September. In a letter addressed to Minister Bertin dated September 7, 1771,[18] Bourgelat wrote:

Signature of Claude Bourgelat

I am pleased that Monseigneur has seen fit to approve our actions taken under the singular circumstances in which we now find ourselves.

Suddenly Mr. Fragonard demonstrated to me what I have always thought to be true: namely, that hatred, like love, can drive one to madness.

Paris le 7 7bre 1771

Mr. parent

Je suis charmé que Monseigneur ait daigné
approuver tout ce que nous avons fait dans
la circonstance singuliere ou nous nous sommes
trouvé. Tout a coup le Sieur fragonard m'a
démontré ce que j'avois toujours cru pouvoir
être, c'est que la haine peut aussi que
l'amour conduire a la folie. le pauvre
homme est en effet fou dans toute la
rigueur et dans toute l'étendüe du terme,
Et tout ce qu'il a dit, tout ce qu'il a
fait depuis huit jours en est une
preuve suffisante. il est inutile d'entrer
icy dans du detail que Monseigneur
connoit et a entendu dire de la bouche
de M. parent. il doit sentir qu'il
n'étoit pas possible de garder un
homme qui 1° vouloit se retirer et
qui 2° souffloit l'esprit de la discorde
dans l'école depuis plus d'une année
sans s'occuper un seul moment de
sa besogne soit en qualité de
directeur, soit en qualité de
démonstrateur. c'est a luy que nous

The poor man can indeed be called mad with all the severity and scope of that term; all that he has said and all that he has done for the last eight days is in itself sufficient proof. It is not necessary to enter here into the details that Monseigneur knows and has heard from the lips of Mr. Parent. It would seem impossible to retain a man who 1) wishes to leave and 2) has spread a spirit of discord in the school for more than a year during which he has not occupied himself for a single moment with his work....

Careful not to contradict too openly the minister's sentiments toward Fragonard, Bourgelat claimed the health of his director had been impaired for some time "by a stone in the kidneys and bladder," preventing him from carrying out his job. He proposed Fragonard be given a pension of 600 livres a year "in gratitude for his previous services," because "we must not, however, abandon him," the letter went on. "Above all he is honest; he has always come to the aid of his sisters and his family, and we should now bestow upon him that same aid. All that I ask of Monseigneur is to have pity on a poor bewildered soul who comes to Paris from the warm sun of Provence...."

Fragonard's pension, raised to 1,000 livres, was in effect until at least 1780. Dismayed upon learning of their director's departure, the pupils expressed their displeasure, leading Bourgelat to convene a meeting at Alfort on September 6, 1771, to calm things down. Bourgelat's announcement of the nomination of Chabert as Fragonard's successor met with immediate applause, so highly was the incoming director regarded by the students.

FRAGONARD'S LIFE AFTER HIS DEPARTURE FROM THE SCHOOL

With his dismissal Fragonard now found himself without home or employment. Yet the revered anatomist experienced no trouble earning a living in an age when beautiful specimens enjoyed brisk sales.[19] Imbued with the spirit of the Enlightenment, a good number of wealthy citizens of worldly ambition were eager to possess their own cabinet of curiosities. Desirous of unusual specimens, they sought to collect not for their anatomical edification but to satisfy their vanity. Fragonard had to work for some of these illustrious amateurs, whom he had perhaps known during his time at Alfort, as well as for the contacts of Minister Bertin, as is attested at the bottom of the last page of his 1792 letter addressed to the national assembly, which mentions "Mr. Fragonard, creator of the cabinet of Alfort near Charenton and of the most valuable specimens in various collections in Paris."

Fragonard's well-salaried directorship combined, no doubt, with the sale of preserved specimens allowed him to accumulate a small fortune,

Letter of September 7, 1771, from Claude Bourgelat to Minister Bertin, in which he recounted Fragonard's dismissal and the ensuing unrest. Archives départementales du Val-de-Marne, Créteil

such that he was untroubled by financial concerns and was able in 1776 to lend the considerable sum of 15,000 livres to Guillaume Robert le Chevalier, Lord of Grèges, conseiller au parlement de Rouen (member of the parliament of Rouen), and a real swindler. Interminable legal procedures followed until 1792, and doubtless Fragonard never saw his money again.[20]

Among notarized deeds sifted through by Pierre-Louis Verly, mainly concerned with this long and unhappy affair, there begins to appear next to the name of Fragonard that of Marie-Anne-Françoise Devaux, widow of François Lafaye, inspector in the office of the captain of the royal hunt in Vincennes. They moved in together in 1778 in the rue de la Mortellerie in Paris in the Hotel de Ville district. They married in 1780, but Marie died three years later.

For eleven years, from 1781 through 1791, historical records are silent: not a single document, not one notarized deed reveals the slightest indication of Fragonard's activities in Paris. All that is known is that the sale of a large house at Carrières de Charenton was made in 1793 to Fragonard by his cousin the painter Jean-Honoré Fragonard. Whether he ever lived there is unknown. It is, however, known that Fragonard did reside for a time at 27 rue de la Tissanderie, at the Hotel des Trois Couronnes, the present-day site of the Bazar de l'Hôtel de Ville, remaining there until his appointment to the Paris School of Health, after which he lodged in the rue de l'Observance until his death in 1799.

It is clear, however, that Fragonard carried on with his anatomical work uninterrupted. The celebrated physician and future director of the School of Health Michel-Augustin Thouret (1748–1810) knew him very well in this time, and he remarked after Fragonard's death that he had carried on working "until his strength was exhausted."[21] A letter of July 1792 reveals that Fragonard was able at that time to offer 1,500 specimens, dried or injected, to the national assembly for the creation of a national anatomy museum. The items that remained in his hands were but a small part of the total he had created, the majority of which was in private collections. Anatomical specimen production was a very lively business indeed!

The procurement of human cadavers before the Revolution was tremendously difficult, at a time when human dissection performed in the amphitheater faced public hostility and religious condemnation; as David Le Breton has described,[22] surgeons, anatomists, and students had to bribe gravediggers in the cemeteries to let them dig up by night the bodies that had been buried the day before. Executioners on the gallows as well handed over the corpses of the hanged in exchange for hard cash. Anatomy could not be learned without "getting your hands dirty,"[23] according to the

Detail from Man with a Mandible

esteemed physician Jacques-René Tenon (1724–1816), risking danger and disgrace, deep displeasure, and serious hazards to one's health and even to one's life. The other source of bodies was the hospital, where many corpses went unclaimed. The skillful Fragonard must have harvested the greatest number of specimens possible from each single cadaver. Evidence shows that he mainly injected and dehydrated organs and body parts separate from the body, such as the head, an arm, the duodenum, and so on, in pursuit of pieces both interesting and useful to study. He surely must have called frequently on the great surgeons' hospitals in Paris. Certainly the beauty

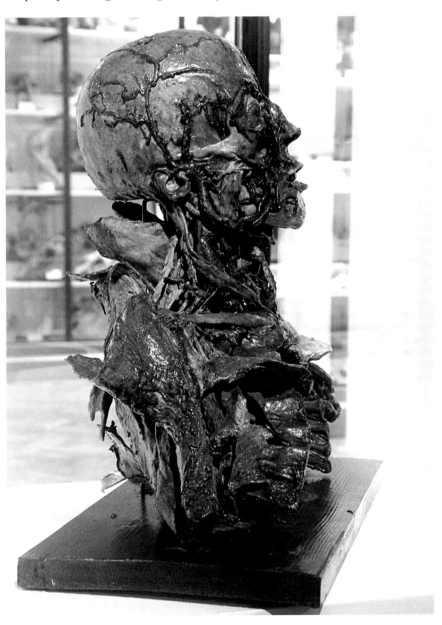

Honoré Fragonard,
Human Bust, no. 1161
in the 1794 inventory.
ENVA, Fragonard Museum

and precision of his works would have attracted their attention and, indeed, when founding the Paris School of Health, they called upon Fragonard.

THE VETERINARY SCHOOL OF ALFORT AFTER FRAGONARD'S DEPARTURE: FROM THE SPLENDOR OF THE 1780S TO THE MISERY OF THE REVOLUTION

The École royale experienced a period of dazzling prosperity, but it was to be short-lived. His health fragile, Bourgelat lived only another eight years after effecting his collaborator Fragonard's dismissal. The position of inspector general of the veterinary schools vacant upon his death, in 1779 at age sixty-seven, was filled by Chabert. Pierre Flandrin (1752–1796), Chabert's nephew, then director of the school at Lyon, was recalled to Alfort to serve as the school's chief director.

As contentious as Bourgelat's personality may have been, the importance of the work he accomplished is indisputable. He left behind two veterinary schools, which through clever publicity aimed at disseminating their founder's good reputation abroad had become the model for other such establishments in Europe. Bourgelat was credited with voluminous scientific production, educational publications, and, with an iron hand, the training of hundreds of students who in turn further spread his agency.

A year after Bourgelat's demise, Henri Bertin stepped down from political life, and in 1781 Louis-Bénigne-François Bertier de Sauvigny was put in charge of agriculture (which included the veterinary schools), ushering in an era of prosperity. Eminent men from the Royal Society of Medicine, the Royal Academy of Surgery, and the Academy of Sciences took an interest in the veterinary school. Daubenton and the anatomist and later celebrated physician Félix Vicq d'Azyr (1748–1794) persuaded the highest state authorities that the school at Alfort should not be limited to education in the art of curing animals, but that it should become a vital center of important scientific studies. Never the master of the fate of his establishment, Chabert now had to accommodate new directions and implement directives from superiors who created a chair in "comparative anatomy" for Vicq d'Azyr and a chair in the "natural history of animals and the rural veterinary economy" for Daubenton. Soon they were joined by Antoine-François Fourcroy (1755–1809), the well-known chemist, who took over chemistry and therapeutics classes. Thus two parallel courses of study were set up at the École d'Alfort: the first, led by veterinary teachers from the school, for students who were to receive only a very basic education; the second, more advanced, was conducted by leading authorities and academics and targeted mainly those

Portrait de Félix Vicq d'Azyr, 1908.
A. Railliet and L. Moulé, *Histoire de l'École d'Alfort* (History of the Veterinary School of Alfort) p. 61., fig. 10, ENVA library

students desiring to expand their knowledge to include biology, which distanced them from the study of veterinary science and from medical and surgical practices of the day on domestic animals.

In November 1783 veterinary training became the responsibility of Charles Alexandre de Calonne (1734–1802), comptroller general of finances, and he assigned to it an extravagant budget. Staff increases and building construction and refurbishments were undertaken. This prosperity was short-lived, however, and the prerevolutionary period saw the imposition of severe restrictions. Calonne's dismissal in April 1787, after four years, resulted in drastic budget cuts, unsettling the comfortable world of the administrators and academics and eliminating the special chairs in 1788. So ended the brief shining moment when the school at Alfort enjoyed a prosperity hitherto unknown, and when it first evolved into a scientific institution.

The Revolution was one of the darkest chapters in the history of the school of Alfort—it was incessantly attacked by those seeking either its closure or its relocation, and its survival is attributable to the determination of the students and to the increasing need, recognized by the elite, for more trained veterinary surgeons. Its former director as well targeted the school in its weakness: stirred by the Revolution, Fragonard aspired to create a national anatomy museum built around the collection of specimens that he himself had created twenty and more years earlier.

FRAGONARD DURING THE REVOLUTION: FROM THE IDEA FOR THE NATIONAL ANATOMY MUSEUM TO THE INVENTORY OF THE COLLECTION AT ALFORT IN 1794

The revolutionary period was incomparably better for Honoré Fragonard: he could at last avenge his ousting and the humiliations he had suffered under Claude Bourgelat's employ.

Fragonard's name is signed to a July 1792 report addressed to the national assembly, in which Fragonard and his collaborators, Delzeuzes and Landrieux, proposed the creation of a national anatomy museum, where the physicians and surgeons badly needed by a nation at war could receive training. They requested to be assigned the positions of directeur, préparateur, and secrétaire, each to be annual appointments, with strict equality of salary at 2,400 livres. In exchange for appointment to these official positions, they offered the nation some 1,500 specimens already in their possession. They requested that the dome of the Church of the Assumption be allocated for the museum, and they envisaged transforming it at very modest cost to include an amphitheater, the collection itself, a preparation laboratory, and lodgings for the employees. Their ambition was to create a "Temple of Anatomy," a virtual cadaver factory, where a great number of specimens would be produced and sold, ensuring both the renewal of the collection and its continued financial support. Although it was sent by the assembly to the comité d'instruction publique (board of public education) on July 22, the proposal appears not to have been followed up. Yet within it were the very ideas that led to the creation in 1795 of a cabinet d'anatomie at the Paris School of Health. The idea for a national anatomy museum was proposed again in 1794 by the gifted wax modeler André-Pierre Pinson (1746–1828), who similarly offered up his collection of anatomical models to be joined by the impressive collection of the duke of Orléans to form a national collection of anatomical models.

Publication of the project by the organisation de l'instruction publique (organization for public education) under philosopher and political scientist Nicolas de Condorcet (1743–1794) prompted Fragonard to submit a supplement to his report to the assembly, in which he attempted to demonstrate the value of his project within the framework of this legislative movement.

Jacques-Louis David (1748–1825), *Self-portrait*, 1794. Musée du Louvre, Paris

Although his idea did not meet with success, it must have caught the attention of the revolutionary authorities, and the painter Jacques-Louis David nominated Fragonard on November 15, 1793, to the Jury national des arts (national board of examiners of the arts). This board was intended to replace the recently dissolved academies and was to award prizes and incentives to patriotic artists while liberating the arts from the yoke of the ancien régime. Among the fifty-five members, experts representing every field of the arts and sciences, were the two Fragonard cousins, Jean-Honoré selected as painter and Honoré as anatomist.[24] As an appointed member, Honoré now had greater authority than his brilliant one-time pupil Vicq d'Azyr, the former physician to Marie Antoinette, who was a deputy of the board.

Of key concern to the Revolution, which had spawned the idea, was the preservation of the national heritage. A vast system of the confiscation and sale of property to profit the nation was put into place. At the close of 1789 the Constituante began selling off crown assets as well as taking over the property of the clergy; in 1792 the national assembly and convention added the confiscation of the property of immigrants. The sale of these goods was initially intended to fill the coffers of the state, but enlightened minds soon recognized the need not only to safeguard their more interesting monuments but, furthermore, to collect any personal objects of educational value for the general populace. The commission conservatrice des monuments (commission for the preservation of monuments), created in December 1790, was the first of the commissions established to achieve these goals. It was charged with the preparation of directives to be sent to

Detail from Fragonard's handwritten 1794 inventory. National Archives, Paris

all départements and was to make an inventory of all religious institutions. As of August 13, 1793, it was joined by a commission of the arts, which was to "catalogue the holdings of the aristocratic academies dispersed in different locations, including all stores of machinery, maps, plans, manuscripts, and other objects of art and science, […] and to ensure that for their conservation these effects be gathered in the one location."[25] Thirty-six commissioners were appointed: among them were, for anatomy, Jean-Baptiste Thillaye (1752–1822), Antoine Portal (1742–1832), principal physician to the king, Vicq d'Azyr, and Jean-Nicolas Corvisart (1755–1821), future primary physician to Napoleon. The commission for the preservation of monuments was dissolved four months later, and the new commission, now at the helm of these important reforms, achieved immediate and impressive results.

On January 14, 1794, Vicq d'Azyr sent copies of his decisive "Directive for the Cataloguing and Conservation, Throughout the Republic, of All Objects Useful to the Arts, Sciences, and Education, Proposed by the Temporary Commission of the Arts, and Adopted by the Committee for Public Education of the National Convention."

Crucial in formulating the concept of a national heritage, this directive formalized the method and principles to be used in creating a general inventory across all areas, with an alphanumeric code to be assigned to each object and associated with a catalogue description of each collection.[26] This essential methodological advance not only laid out the procedures for the inventory of objects but also designated all personnel required by the temporary committee for the arts to carry out the task of cataloging.

Along with these guidelines, Vicq d'Azyr's directive listed the collections and stores to be inventoried by the commission of the arts, which included the Veterinary School of Alfort. Clearly identified as a collection important to national heritage, the specimens at Alfort were to be meticulously inventoried before their removal would begin. Included in the board of public education's list of members who were to "inventory and properly house books, instruments, machinery, and other objects of science and the arts" were, "for the inventory of collections of anatomy, the citizens Thillaye, anatomist Fragonard, Vicq d'Azyr, Corvisart, Portal."[27] And so it was that Fragonard was reunited with the celebrated anatomists of the end of the century, and that the doors of the collection at Alfort were once again opened before his eyes.

Having been forewarned about specimens having disappeared from the collections at Alfort, on January 3, 1794, Thillaye visited the school to inspect the collection with the director, Pierre Flandrin. He noted that, despite the poor condition of the premises, the specimens were well preserved, perfectly classified, and inventoried and described by the pupils themselves.[28] Four months later, however, a report of May 14, 1794, from Fragonard provided alarming news about the neglect of the premises and collections at Alfort

Inventory label affixed in 1794 to the shoulder blade of a dolphin. The label II D 598 conforms to the format of Vicq d'Azyr's specifications: *II* indicates the Alfort collection, *D* anatomy, and 598 is the number of the object.

at the end of that winter, and on May 24, 1794, he advised its relocation to Paris, encouraging the creation of his national anatomy museum.

> That the cabinet of anatomy at the national Veterinary School of Alfort near Charenton be conserved is all the more imperative in that it is the only one in the entire Republic that can serve in the study of comparative anatomy after the convention has organized educational discipline.… It seems unwise for a collection of such import to remain in the long term so far from Paris, where the training of comparative anatomy ought to be established, and outside of which only a very small part of the intrinsic value of such a collection can be taken advantage of.[29]

Following the principles established by Vicq d'Azyr, Fragonard now also drew up a detailed inventory (countersigned by Corvisart)[30] enumerating the estimated 3,033 specimens stored in the collection at Alfort. Thorough and precise, the inventory listed each item (duly provided with a label inscribed with a unique number) with a brief description and sometimes remarks on the preservation technique used or the specimen's condition. All together it filled 186 folio pages in a very oversized book. The book was deposited with the temporary commission of the arts on June 28, 1794, and officially accepted on January 13, 1795, by Pierre Flandrin, the director of Alfort; Fragonard; and two other members of the anatomy section, le Clerc and Thillaye. Some of the oldest specimens in the Fragonard Museum today bear the marks of that inventory: visible on each of the *écorchés* is a number written in red paint, probably from the earliest days of the cabinet.

In April 1794 Fragonard was also in charge of the appraisal of the cabinet of "ci-devant Orléans" (aristocrat Orléans) in the Palais Royal. The following July he was entrusted with the examination of the very substantial anatomy and zoology collection of the renowned professor of anatomy in Paris Jean-Joseph Sue the younger (1760–1830), the son of Jean-Joseph Sue *père* (1710–1792), the author of *Anthropotomie ou l'art d'injecter* (Anthropotomy, or The Art of Injection), who likely had been an inspiration to Fragonard in the art of preparing and conserving bodies. Fragonard collaborated the same year too in the inventory and assessment of the collection of the former Academy of Surgery, and of the former faculty of medicine of Paris the next year. With Thillaye he inventoried as well the bones from the Cimetière des Innocents (Cemetery of the Innocents) and traveled to Toulouse to inspect the bodies in the Franciscan cellars.

In this same year Thillaye became head of specimens of the new Paris School of Health created by an act of December 4, 1794. This school was intended not only for training medical officers in the army and the military hospitals but also in some respects as a replacement for the recently dissolved faculties of medicine and surgery. On March 4, 1795, the board of public education issued a decree concerning the staff of the schools of health whose work it was to prepare anatomical specimens.[31] The decree placed Fragonard, with a salary of 5,000 livres, "in charge of directing research and anatomical preparation and training students in the art of injections."[32]

A new decree just eight days later from the board confirmed his nomination as chef des travaux anatomiques (chief of anatomical works); it also named the six prosectors who would join him: Constant Duméril (1774–1860), who became a zoologist and celebrated physician; Guillaume Dupuytren (1777–1835), future eminent surgeon and physician to Charles X; Pierre Desauge; Dufau; Lassis; and Ribes.[33] In addition to these prosectors the staff was to include none other than the famed painter and modeler in wax André-Pierre Pinson, thus bringing together the two most important figures working with anatomical preparations, the masters, respectively, of natural and artificial anatomy. Pinson, who had extensive surgical experience and had skillfully prepared wax anatomical models for the celebrated collection of the former duke of Orléans, was now to devote himself to creating models of lesions based on dissection of pathological specimens, the area of practice in which Dupuytren was to flourish so spectacularly.

A small, well-documented volume by Michel-Augustin Thouret, first director of the Paris School of Health, found by Verly in the Bibliothèque nationale, provides a rare record of Fragonard's activities there.[34]

> To convey the magnitude and invaluableness of the school, one need only point to what was done in anatomy. Since the founding of the school, at least three hundred students have been enrolled each winter to participate in dissection courses. Each year the cadavers used for this area of instruction numbered upward of three or four hundred....
>
> The formation of the cabinets of the school is a long-term endeavor; the formation of six collections of anatomical specimens, pathological specimens,... etc., calls for a lifetime's exclusive devotion to this work....
>
> Three artists are engaged in the preparation of these anatomy collections and those of other schools: the first is citizen Fragonard, to whom we are indebted for the fine collection of anatomical pieces at the Veterinary School of Alfort; second is citizen Pinson, wax modeler; third is citizen Lemonnier, the painter and draftsman, whose illustrations grace the descriptive catalogue of the museum.

For their abilities as teachers no less than for the perfection of their art, these three artists are equally vital; the creation of the school's collections depends entirely on them....

The formation of the cabinets of collection has begun with much zeal and met great success. Already to be admired is a series on surgery, the most comprehensive to date in Europe. Also on view are a great number of anatomical pieces natural and artificial, a valuable collection of materia medica, of diseased bones, pathological specimens modeled in wax, and so on.

Less than two months after the creation of the Paris School of Health, in a faculty meeting convened on January 23, 1795, Thillaye was entrusted with the organization of a collection of natural history and comparative anatomy, which was to include items of interest from the cabinet at Alfort. Thillaye viewed the Alfort collection with a fresh eye. In April and May he produced a list of the specimens best suited to his project. The board for public education issued a decree dated May 29, 1795, concerning the proposed transfer, but the transfer of the Alfort specimens to the museum of the Paris School of Health did not take place until July 10, 1795. The list of objects transferred has to this day never been found, making it impossible to know exactly which items were removed; what's more, the museum was also permitted at the time to take specimens of animal anatomy.

There is but one reference extant, from Paris School of Health director Thouret, dated June 9, 1795, which mentions the items transferred numbering between 1,000 and 1,200 pieces, a third of the collection:[35] "A thousand to twelve hundred anatomical specimens of human anatomy were taken from the superb collection of the Veterinary School of Alfort," and "these created the foundation collection at the School of Health, to which were added specimens from the cabinet de l'Académie des sciences, the Faculté de Médecine, and the hospice de l'Unité[36] as well as twice as many more offered by the Muséum d'Histoire Naturelle."

The removal of so substantial a number of specimens coupled with the deterioration of the collections drastically reduced the value of the prestigious cabinet at Alfort that Fragonard had built up. But the school was in no position to resist its plundering as it struggled just to survive. The new collection that had been gathered constituted the nucleus of a world-renowned collection that came in the twentieth century to be known as the Musée d'Anatomie Delmas-Orfila-Rouvière, located on the top floor of a building within the Faculty of Medicine at the Paris V René Descartes University in the rue des Saint-Pères. With some 5,500

Detail from *Human Arm*

specimens, the Musée Orfila, as it was commonly called, contained, along with the Anatomy Museum of the faculty of Montpelier, what was perhaps the most beautiful collection in France. The collections of the Musée Orfila have been placed into storage and are no longer viewable; the museum at Montpellier is open only occasionally.

THE END OF FRAGONARD'S LIFE, AND HIS LEGACY

After December 1795 Fragonard no longer attended the sessions of the temporary commission of the arts that he had most diligently participated in, and he concentrated exclusively on his new post as director of anatomical research.

Four years after his appointment, Fragonard died on April 5, 1799, at age sixty-six, at his home in the rue de l'Observance, after a brief illness. Two employees of the School of Health witnessed his last hours. Thouret called a meeting of the faculty three days later to announce and discuss the circumstances of his death and to eulogize his colleague.

Citizen Thouret announced the circumstances of the illness leading to the death of citizen Fragornard [*sic*], chief of anatomical studies at the school.

The assembly deeply regrets the loss of one of its most dedicated collaborators whose reputation in the art to which he devoted his life is known among scholars and indeed all who are engaged in research and the pursuit of the understanding of anatomy. Simple, humble, and eschewing anything approaching not only luxury, but even mere immoderation, Fragornard [*sic*] lived apart from high society. Should he be required to put in an appearance there from time to time, he would abandon it the instant it demanded of him behavior that conflicted with his simple and the plain morals. One incident, for example, that illustrates the nobility and strength of his character: Minister Bertin, under whose jurisdiction the veterinary school was placed, greatly admired Fragornard's [*sic*] expertise and invited him to dine at his table.

The anatomist arrived wearing attire decent enough but poor and certainly not of the refinement befitting a meeting with the gentry to whom he now presented himself. The minister's brother, the abbé Bertin, was compelled to offer to cover the expense of a new suit. The next day fabric for the tailoring of a complete suit was sent to and refused by Fragornard [*sic*], who sent it back and never again came to call at the house of the minster.

Wholly devoted to his anatomical work, with imagination Fragonard created new specimens he was so resolutely determined should be seen, but he wrote nothing. As was said of Rueschs [Ruysch], it can be said of him, that he communicated with our eyes. To all who inquired about his collection was his reply: *Come and see.*

If the school was not able to take advantage of Fragonard's talents as much as might have been hoped, due to his exhaustion from his industrious preparation of so many specimens already before he came to the school, surely also to blame are the successive losses he suffered in his fortunes, which affected his character, making him less communicative, though his passion for anatomy remained strong as ever. Citizen Dubois, who opened his body for autopsy, discovered a tumor occupying the duodenum, part of the pylorus, and the liver, and has promised to give a written account, which the school may find fit to attach to the present verbal report.[37]

Seven candidates vied to succeed Fragonard: Duméril, Giraud, Marie-François-Xavier Bichat (1771–1802), Dupuytren, Nicolas Jadelot, Valentin, and Dominique-Jean Larrey (1766–1842). That such illustrious physicians and scholars sought to assume his position reflects the importance of the post he had held. The final competition came down to Duméril and Dupuytren, and it was the former who was awarded the position. Duméril later became professor of anatomy, in 1801, and Dupuytren then took over as head of anatomical preparations.[38]

Honoré Fragonard is one of the rare anatomists remembered not through the written word but by his actual art, the very essence of anatomy: dissected and preserved anatomical specimens. While history has made the full extent of his contribution to the evolution of understanding the science of anatomy impossible to know, he was beyond doubt an anatomist of genius, a practitioner unequaled. The degree of his skill was so high, it could come only from having fully mastered the architecture of the body and its structures revealed by the scalpel. He was key in the century in which anatomical study became academic and attained precision and in which the anatomical *corpus* obtained a completeness prior to the great upheavals of the nineteenth century. The gross anatomy of man was henceforth known, and as normal anatomy consequently lost significance for research two new disciplines arose: physiology and pathological anatomy. Physiology looked at the relationship between organ and function; the anatomist described not only the organ's morphology but its working within the overall anatomical structure. But now anatomy began to focus on the study of lesions caused by disease, and the body became the terrain of exploration into pathophysiological disease. Just as the science of anatomy was changing, fittingly so too was the work of Fragonard being superseded by the work of Dupuytren, the pioneer of anatomical pathology.

The name Honoré Fragonard will always be associated with the specialized field of the art of injection in anatomical preparation. His was the century in which the creativity of anatomical scientists reached its highest peak. He was also a pioneer in animal anatomy, the first to teach what was an almost unknown discipline, formalizing the study of the lowly species and illuminating the impact of the animal kingdom on the nation's economic development. The work had been arduous but results were dazzling. The collection Fragonard founded at Alfort was the first great collection of comparative anatomy in the country, and its renown spread throughout the world far beyond the borders of France.

AN·ÆT·XXVIII
M·D·XLII

From the Royal Veterinary College to Modern-Day Plastination

"Come and see"

Fragonard was born nearly two centuries after the publication in 1543 of Andreas Vesalius's revolutionary work, *De humani corporis fabrica libri septem*. In Italy in the sixteenth century the universities of Padua and Bologna in particular welcomed the renaissance of anatomy, vigorously dissecting to rectify the many errors in anatomical understanding that had remained sacrosanct for centuries. The prevailing theories were turned upside down, and the way the human body functions was completely rethought. The *Fabrica* described anatomy in detail and irreversibly changed the status of the anatomist. Previously, the learned physician had been an interpreter of ancient texts, a philosopher who did not confront the mortal remains himself; rather, it was the unqualified prosector who carried out the thankless task of dissection, which was supposed to confirm in the flesh the assertions of the masters of antiquity. Vesalius, on the contrary, performed his own dissections, plunging his own fingers into the foul body fluids to extract new truths. He no longer taught based solely on an academic text, but used illustrations to make a visual impression on his pupils. He recorded on paper exactly what he himself had observed, and thus anatomy began to distance itself from its ancient tradition, which was so often at odds with the physical facts.

This rejection of the old sources rapidly began to spread while the use of vernacular language ensured the communication of a new understanding in the emerging fields of medicine and surgery. As ideas about gross anatomy were becoming formalized, the anatomists of the seventeenth century were able to make the discoveries that revolutionized knowledge of human physiology. The circulation of the blood, for example, was identified in 1616 by William Harvey (1578–1657),[1] and the lymphatic system in 1622 by Gaspare Aselli (c. 1581–1626).[2] The development of the microscope made minute structures visible, enabling Marcello Malpighi (1628–1694) to discover capillaries in 1661. The anatomist now searched to discover the link between the

Portrait of Andreas Vesalius, De humani corporis fabrica, Basel: Joannes Oporinus, 1555. BIUM, Paris

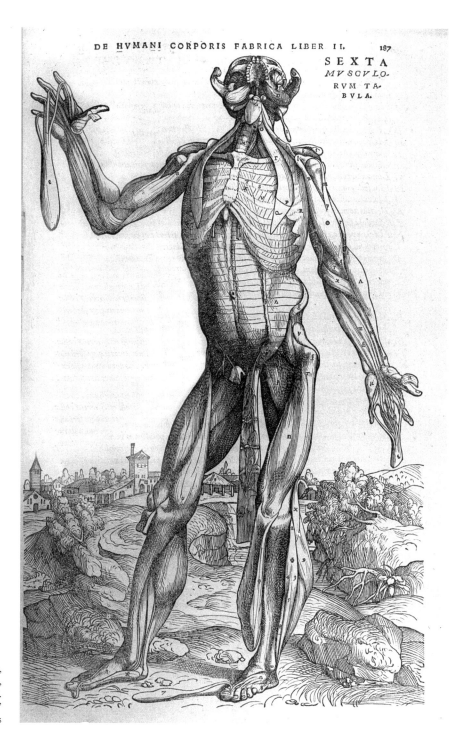

Andreas Vesalius,
De humani corporis fabrica,
Basel: Joannes Oporinus, 1543.
"Sexta musculorum tabula…"
BIUM, Paris

FIGVRA HVIC CHARTAE IMPRESSA, CVIQVE ALIA VARIIS

FORMATA PARTIBVS AGGLVTINATVR, ORDINE FIGVRARVM MVSCVLOS OSTENDENTIVM

quinta numerari potest. eos enim proponit, qui ex omnibus in anterio
ri sacie uidendi sunt reliqui. Verùm quum hi paucissi
mi sint, non mirum est, præsentem figuram magna ex parte, uti &
quæ sinistra continetur manu, illi succedit quæ quartæ sini
media occurrit pars) iacet. quod uerò in caluariæ amplitudine spe
ri subijciuntur. Vbi etiam figura cum suis partibus explicatur, quæ
uiris peculiaria exceperis, præsenti pagina etiam indicanda.

præcedentem, nuda ossa proponere. Cerebri figura,
stra amplectitur, & præsenti illa quæ humi (ubi oculi
ctandu est reliqui, illis uidendum est figuris, quæ mulie
præsenti figuræ agglutinatam cernis, si modò organa

FIGVRAE HVIC CHARTAE IM
pressæ characterum Index, una cum eorum cha
racterum explicatione, qui organis generationi sa
mulantibus in figura presenti chartæ
agglutinata inscribuntur.

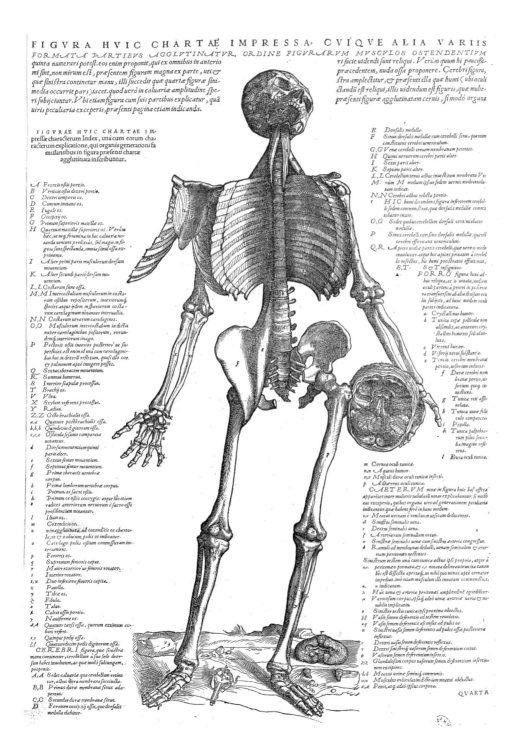

A Frontis ossis portio.
B Verticis ossis dextri portio.
C Dextri temporis os.
D Cuneum imitans os.
E Iugale os.
F Occipitij os.
G Primum superioris maxillæ os.
H Quartum maxillæ superioris os. Verùm
bæc, ut uis foramina in hac caluaria no
tanda ueniunt prolixiùs, sed magis in sin
guos sunt spectanda, omnia simul ossa ex
primente.
I Alter primi paris musculorum dorsum
mouentium.
K Alter secundi paris dorsum mo
uentium.
L, L Costarum sunt ossa.
M, M Intercostalium musculorum in costa
rum ossibus repositorum, interiorumq́;
species. atque idem in superiorum costa
rum cartilaginum notantur interuallis.
N, N Costarum ueruarum cartilagines.
O, O Musculorum intercostalium in dictis
nuber cartilaginibus positorum, eorum
demq́; interiorum imago.
P Pectoris ossa interior posterior' ue su
perficies. est enim id una cum cartilagini
bus huc in dextrū reflexum, quæsi illo cor
& pulmonem apté integere posset.
Q R Sextus thoracem mouentium.
S Summus humerus.
S Interior scapulæ processus.
T Brachij os.
V Vlna.
X Stylum referens processus.
Y Radius.
Z, Z Octo brachialis ossa.
a, a Quatuor postbrachialis ossa.
b, b, b Quindecim digitorum ossa.
c, c, c Ossicula sesamo comparata
notantur.
d Dorsum mouentium quinti
paris alter.
e Sextus femur mouentium.
f Septimus femur mouentium.
g Primæ thoracis uertebræ
corpus.
h Primæ lumborum uertebræ corpus.
i Primum os sacri ossis.
k Primum os ossis coccygis: atque hæc etiam
radices anteriorum neruorum è sacro osse
prosilientem notantur.
l Ilium os.
m Coxendicis os.
n in aggutinatione ad coxendicis os charte
le, ut & o obolium, pubis os indicatur.
o Cartilago pubis ossium commissuram in
teruenies.
p Femoris os.
q Supremum femoris caput.
r Maior exterior' ue femoris rotator.
s Interior rotator.
s, u Duo inferiora femoris capita.
x Patella.
y Tibiæ os.
z Fibula.
& Talus.
ſ Calcis ossis portio.
? Nauisorme os.
A, A Quatuor tarsi ossa, quorum eximum os
hunc refert.
t, t Quinque pedij ossa.
l, l Quatuordecim pedis digitorum ossa.

CEREBRI figura, quæ Delineatio
menta continerunt, cerebellum à sua sede deor
sum habet inuolutum, ac quæ modò subtungam,
proponit.
A, A Sedes caluariæ quæ cerebello retine
tur, atque dura membrana succindat.
B, B Primæ duræ membranæ sinus ada
pertus.
C, C Secunda duræ membranæ sinus.
D Foramen occipitij ossis, quo dorsalis
medulla elabitur.

E Dorsalis medulla.
F Sinus dorsalis medullæ cum cerebelli sinu, quorum
constituens cerebri uentriculum.
G, G Vrnæ cerebelli tenuem membranam petentes.
H Quinti neruorum cerebri paris alter.
I Sexti paris alter.
K Septimi paris alter.
L, L Cerebellum tenui adhuc inuestitum membrana. V,
M. tùm M mediam ipsius sedem uermis modo inuolu
tam indicat.
N, N Cerebelli adhuc relicta portio.
r HIC humi decumbens figura inferiorem cerebel
li sedem commonstrat, quæ dorsalis medullæ commis
tebatur initio.
O, O Sedes quibus cerebellum dorsali continuebatur
medulla.
P Sinus cerebelli cum sinu dorsalis medullæ, quarti
cerebri efformant uentriculum.
Q. R Apices mediæ partis cerebelli, quæ uermis modo
inuolutæ, atque hos apices priuatim à cerebel
lo resectos, hic humi prostratos essinxus,
S, T. S T insignios.
■ PORRO figura humi ad
huc reliqua, æ à notatæ, mediam
oculi partem, à priori in posterio
ra transuersim ab alia diuisam ocu
lis subijcis, ad hunc modum oculi
partes indicatura.
a Crystallinus humor.
b Tunica esse pellicalæ non
absimilis, ac anteriori cry
stallinis humoris sedi eluo
lutas.
c Vitreus humor.
d Visorij nerui substantia.
e Tenuis cerebri membranæ
portio, uisorium induces.
f Duræ cerebri mem
branæ portio, ui
sorium quoq; in
uestiens.
g Tunica uuæ assi
milata.
h Tunica uuæ foli
culo comparata.
i Pupilla.
k Tunica palpebra
rum pilos suor
ha imagine refe
rens.
l Dura oculi tunica.

m Cornea oculi tunica.
n, n Aqueus humor.
o, o Musculi duræ oculi tunicæ inserti.
p Adhærens oculi tunica.
CAETERVM nota in figura huic hoc affixa
apparétes inter mulieris tabularú notas explicabuntur: si modò
eas exceperis, quibus organa uiro ad generationem peculiaria
indicentur, quæ habent sere in hunc modum.
v, v Meatus urinam è uenis in uesicam deducentes.
t Sinistra seminalis uena.
s Dextra seminalis uena.
ſ Arteriarum seminalium ortus.
u Sinistræ seminalis uenæ cum sinistra arteria congressus.
x Ramuli ad membranas deducti, uenam seminalem & arte
riam peritonæo nectentes.
Sinistrum testem una cum tunica adhuc ipsi propria, atque à
peritonæo pronata, t, t notata delineauimus: ita tamen
hic est dissecta aperta q́;, ut nihil quo minus apté cernatur
impedias. imò etiam musculum illi innatum communis via,
u indicatus.
H Hæc uena & arteria peritonæi amplitudine egrediütur.
μ Varicosum corpus si suij adeò uenæ arteriæ uería &
nebilis implicatio.
ξ Sinister testis ipsi proxime obiectus.
H Vasis semen deferentis ad testem reuolutio.
e, g Vasis semen deferentis ascensus ad pubis os.
n Sinistri uasis semen deferentis ad pubis ossis posteriora
inflexus.
σ Dextri uasis semen deferentis reflexus.
ρ Dextri sinistriq́; uasorum semen deferentium coitus.
φ Vasorum semen deferentium inserto.
χχ Glandulosum corpus uasorum semen deferentium insertio
nem excipiens.
ψ, ψ Meatus urinæ seminiq́; communis.
ωω Musculus orbiculatim è solo iam meatui obductus.
ε, ε Penis, atq; adeò insius corpora.

QVARTA

Andreas Vesalius, *Andreae Vesalius…suorum de humani corporis fabrica epitome, De humani corporis fabrica,* "Figura huic chartae impressa…" Basel: Joannes Oporinus, 1543. BIUM, Paris

organ and its function—not through mere theory but actual substantiation and proceeding by logical deduction through a complexity that seemed limitless.

One of his era's most eminent representatives, Fragonard made key contributions to the booming field of anatomical science, which in some respects was still something of a social event. Anatomy was nearing its peak of achievement, in both the reach of its knowledge and in conservation and reproduction techniques. Some of the best minds of the century were drawn to anatomy and honored by the new revolutionary government in its committees and institutions dedicated to art and to science. Paradoxically anatomy was to lose its fruitfulness and power in the conquest of scientific knowledge even as it was attaining its peak. But is it not the way of all disciplines that decline begins upon reaching the apex?

From the end of the sixteenth century, as science progressed and the social standing of the surgeon had risen, so too multiplied the activity of dissection. Not only the scholars but a curious public too was keen to be initiated into the mysteries of the human body, and the spectacle of it was all the rage. The universities began offering sessions to the paying public, rivaling the anatomical theaters that had been opened specifically in response to public demand. Posters publicized these dissection sessions well in advance. In the five-hundred-person amphitheater of Louis XIV's Jardin du Roi (now the Muséum d'Histoirie naturelle), built in Paris in 1673, large crowds gathered in a craze that lasted until the Revolution.[3] In 1673, too, Molière took advantage of the trend to entertain his public with *The Imaginary Invalid*[4] (before dying during a production of the same):

> Thomas Diafoirus, again bowing to Angélique: "With the permission of this gentleman, I invite you to come one of these days to amuse yourself by assisting at the dissection of a woman upon whose body I am to give lectures."

> And the servant Toinette, laughing: "The treat will be most welcome. There are some who give the pleasure of seeing a play to their lady-love; but a dissection is much more gallant!"

There had long been a demand for cadavers for surgical training, and that demand was now increased by the curiosity of the public. Several procedures were essential to retard rapid decomposition so that dissected cadavers could be safely kept for the curious, for students, or for the artists

who created anatomical models. Dissected bodies could be preserved by two methods: submersion of organs or body parts in preservative fluids in clear glass containers or dehydration of the tissues and preservation of the body—mummification.

Artificial reproductions of the body or dissected organs could be made as well. The technique of modeling, in wax for human subjects, sometimes in plaster for animals, had spread from Bologna and Florence, and models became a useful addition to anatomical engravings, having the added advantage of three-dimensionality. The number of anatomical waxes still preserved today, made with remarkable skill, shows their unparalleled value in teaching. In wax the artist copied the shape, the size, the colors, and the textures of tissues and organs. Even the invisible structures were made visible; the autonomic nervous system or the lymphatic vessels, for example, were clearly revealed in the waxes of the surgeon Jean-Baptiste Laumonier (1749–1818), who opened a factory of wax modeling in Rouen, which flourished briefly. Another French preparator, and protégé of the duke of Orléans, was André-Pierre Pinson, surgeon to Louis XVI and artist of wax models. Pinson enriched the collection of the Palais Royal with some remarkable works that were later requisitioned for the Museum of Natural History. He had in fact offered his personal collection to the new institution before being appointed to the new Paris School of Health in 1795, the same time as Fragonard.

LEFT: André-Pierre Pinson, *Seated Woman, Anatomy of the Viscera*, c. 1780. Wax. From the collection of the duke of Orléans. Muséum national d'histoire naturelle, laboratoire d'Anatomie comparée, Paris

RIGHT: Jean-Baptiste Laumonier, *Angiology of the Thorax, Neck, and Head.* Wax. Musée Flaubert et d'histoire de la médecine, CHU-Hôpitaux de Rouen

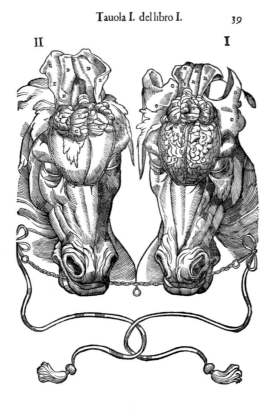

II   I

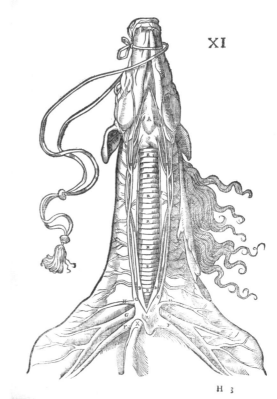

XI

H 3

THE PIONEERS OF A COMPARATIVE ANATOMY DEDICATED TO THE CARE OF ANIMALS

Carlo Ruini, *Anatomia del cavallo infermita et rimedii*. Venice, 1618. Left to right, Book 1, plate 1, figs. 1 and 2: dissections of the brain in situ. Book 2, plate 2, fig. 11: dissection of the ventral cervical region. Book 5, plate 3, fig. 3: arterial system of the horse. Book 5, plate 5, fig. 3: myology of a trotting horse, rear view.

In 1762 Fragonard found himself in what must have been a very uncomfortable situation. Equipped with only his knowledge and experience as a surgeon of humans, he was to begin to lay the foundation of the scientific study of animal anatomy when he joined Claude Bourgelat in Lyon, as a teacher of anatomy in the first veterinary school in France. Existing literature on the subject was summarized in a few works as rare as they were expensive, all concerning the horse.

The oldest, published in 1598, the *Anatomia del cavallo*[5] by Senator Carlo Ruini (c. 1530–1598), was the first treatise describing the anatomy of the horse. Its first part is illustrated by sixty-four full-page plates[6] of stunning beauty—the simplicity of the depiction of anatomy, stripped to the extreme, is in sharp contrast to the delicacy and complexity of the illustration of the mane and tail or of the harness. The use of illustrations in this work, probably inspired by the Vesalius's *Fabrica*, marked a decisive turning point in the study of the anatomy of the horse. Might Fragonard have held this masterly work between his hands? Probably.

The second oldest work is the French *Hippostologie*[7] by Jean Héroard (1551–1628), physician to the future Louis XIII, published in 1599. Jean Héroard, lord of Vaugrigneuse, was known for his detailed *Journal*,[8] in

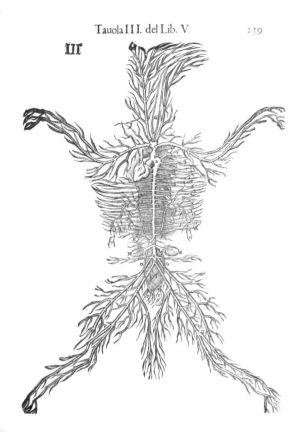

which he recorded the daily life of Louis since his birth. Héroard's *Journal* is filled too with information and fascinating anecdotes on the role of animals in the French courts. With *Hippostologie* Héroard was the first Frenchman to publish a work on the anatomy of the horse, albeit limited to osteology. The seven generally accurate plates depicting the skeleton were reprinted for more than two centuries in works of "empirical" medicine used by the farriers or landowners, in which they were mixed together with old theories of humors and astrology in an astonishing jumble of formulas. Did Fragonard have access to Héroard's document? There is no proof one way or the other.

The veterinary authors of the seventeenth century scarcely mentioned anatomy and, when they did, their descriptions were so vague and erroneous, they clearly had not witnessed a single dissection or even had any texts on the subject to refer to.

Claude Bourgelat was the first important French contributor to the subject in the eighteenth century, publishing as a riding master and director of the l'Académie royale d'équitation de Lyon (Royal Academy of Equitation at Lyon), a dry, precise treatise containing only a single illustration. His *Elémens d'Hippiatrique*[9] was published in two volumes between

LE CORPS DES OS DV CHEVAL

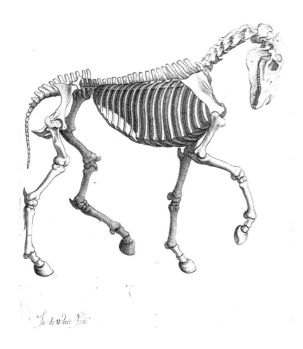

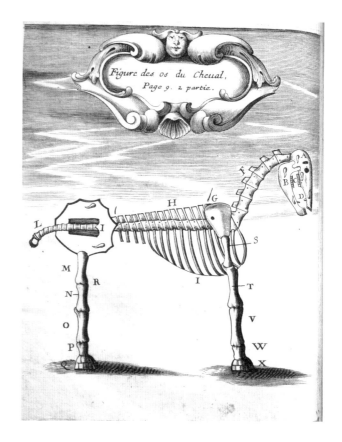

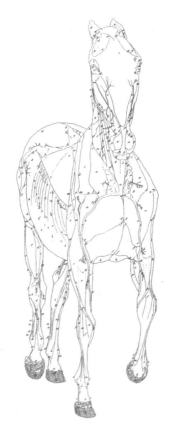

LE CHEVAL.

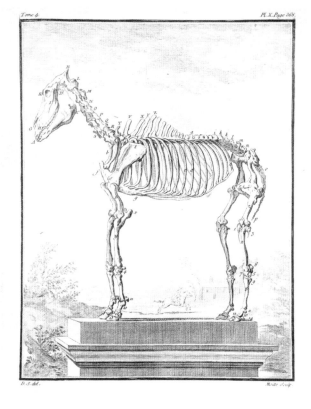

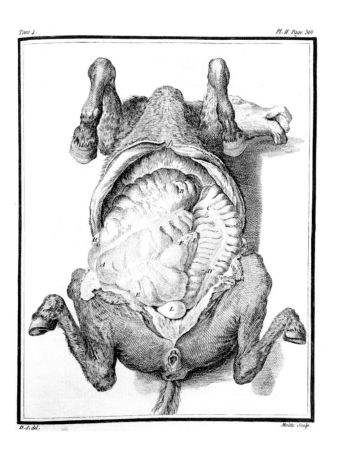

CLOCKWISE FROM TOP LEFT
Georges Leclerc de Buffon, *Histoire naturelle* (Natural History). Paris: imprimerie royale, 1753. The horse, plate 1, p. 212; the skeleton of the horse, plate 10, p. 366; the abdominal viscera of the dissected horse, vol. 4, plate 2, p. 366. ENVA library

1750 and 1753. Within its dialogue of questions and answers between a master (presumably the author) and a pupil are descriptions of the utmost precision demonstrating the quality of Bourgelat's research. The single plate illustration[10] represents the ideal "geometrical proportions" of the horse, a late and rather clumsy adaptation of the Vitruvian Man, Leonardo da Vinci's diagram of human proportions. Certainly Fragonard would have had access to this treatise written by the veterinary school's founder.

At the same time, in 1753, Buffon published the fourth volume in his *Natural History*, the volume in which he described the anatomy of domestic animals. Although the text may not have been as detailed as Bourgelat's, this pioneering work was enhanced by superb engravings depicting a number of organs and predicted the future direction of scientific veterinary medicine.[11]

Detail from *The Anatomy Lesson at the House of Lafosse* (p. 136). A young apprentice demonstrates the anatomy of the horse's hoof to his peers. ENVA

One of the most beautiful works ever produced on equine anatomy, *The Anatomy of the Horse*,[12] was published in 1766 by the English painter George Stubbs (1724–1806). Not strictly a work of anatomy but a collection of eighteen plates showing successive stages in the dissection of the horse,[13] it revealed the deep structures of the animal's conformation.

Lastly, Philippe-Étienne Lafosse (1738–1820), Bourgelat's rival and founder of a competing school, undertook the composition of a monumental work, *Cours d'hippiatrique* (Veterinary Studies), which appeared in 1772. This folio, superb in all respects, brought its author very wide recognition—and ultimately ruination, for the publication of this marvel, with sixty-five plates engraved in color, was at the author's own expense.

It was by carrying out careful dissections with his pupils, not by studying these often incomplete descriptive works, that Fragonard was to master animal anatomy so quickly. Reflecting on the progress of veterinary science, years later in 1792 he wrote, "continually hunched over cadavers [the anatomist] advances in his research, realizes his conjectures; each day brings something new, for in anatomy we do not yet know everything; and like practical medicine, it cannot be learned from books." Pierre-Christian Abidgaard (1740–1801), a pupil at the school in Lyon in 1763 before going on to found the school in Copenhagen, recorded these methods in his *Histoire abregée de l'École de Copenhague*[14]: "The principles of human medicine were adapted to the veterinary art. Cadavers were used in anatomy lessons. Satisfied that his books had been copied out, the school founder Bourgelat passed the responsibility of explaining them to the pupils to two demonstrators in surgery, Pons and Fragonard; the latter was a skillful anatomist, but both were little versed in the veterinary arts."

It was Jean-Baptiste Huzard (1755–1831), director of the school, who in 1798 elucidated Fragonard's vital contribution to the *Précis anatomique du corps du cheval* (Anatomical Handbook of the Body of the Horse), published in 1768 by Claude Bourgelat deliberately without the least mention of Fragonard's work: "Mr. Bourgelat had begun also in 1768 and 1769, jointly with Mr. Fragonard (a highly educated anatomist, to whom the veterinary schools are much indebted), an account of the principal differences observed in the dissection of the ox, the goat, the ram, and their females, in comparison to the horse and the mare."[15]

In a notice that he inserted into the third edition of *L'Art vétérinaire ou médecine des animaux* (Veterinary Arts, or The Medicine of the Animal),[16] probably written by Claude Bourgelat, Huzard again stressed the importance of Fragonard's work: "I cannot allow to pass unnoticed that

Pl. XXXII.

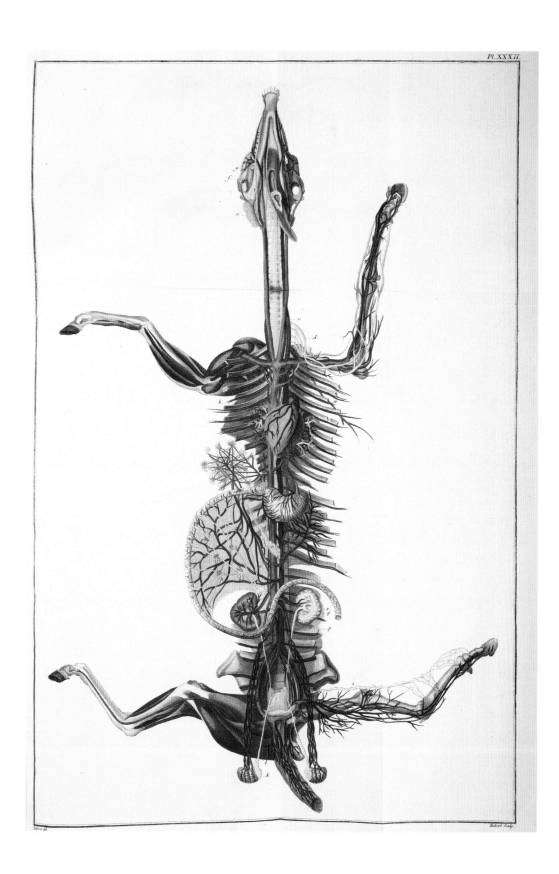

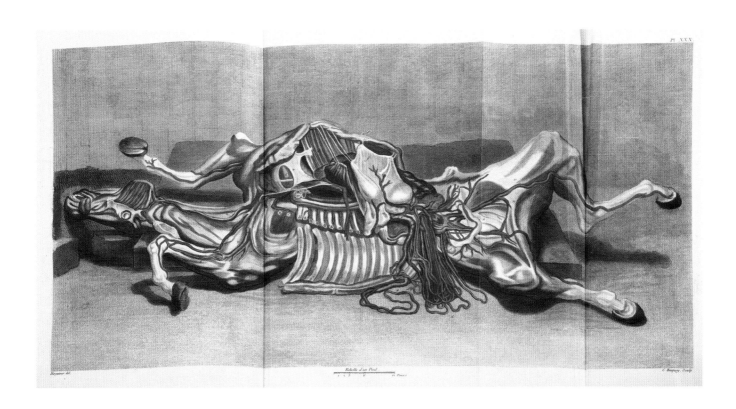

Philippe-Étienne Lafosse, *Cours d'hippiatrique ou Traité complet de la médecine des chevaux* (Veterinary Studies, or A Complete Treatise on Equine Medicine). Paris: Edme, 1772. ENVA library

OPPOSITE: Plate 29 showing a complete angiology

ABOVE: Plate 30 showing the principal venous trunks

Surgeon Fragonard, the distinguished anatomist whose work in injection is as worthy as that of Ruysch, former director of the Veterinary School of Alfort and currently at the Paris Medical School, has undertaken, with Bourgelat, most of the anatomical investigations which resulted in this book." And Jean Girard (1770–1825) added in 1819: "Bourgelat had supplied the rudimentary anatomy of the horse…and, enormously aided by the industrious Fragonard, published in 1768 a complete treatise on this most important area in the veterinary art."[17]

In its diversity and sheer abundance, the collections of the anatomy cabinet Fragonard and his collaborators had pioneered showed how broad their research had been. Three thousand and thirty-three specimens of all different techniques were listed in the 1794 inventory drawn up by Vicq d'Azyr and Fragonard,[18] a topographical description of the collections arranged by room numbers and within each room, by cabinet, table, or display case.

> Inventory of the cabinet of anatomy of the National Veterinary School of Alfort:
>
> 1. Anatomical specimens
> 2. Pathological specimens
> 3. All the shoes and plaster models for horseshoeing
> 4. Plaster models of different animal parts
> 5. Plaster molds
> 6. Anatomical drawings
> 7. Engravings
> 8. Charts related to the veterinary art
> 9. Copper boards for bandages
> 10. Surgical instruments
> 11. Instruments for anatomical injections, including mercury among others.

Because the inventory does not list the creator of each piece, it is impossible to attribute some of the specimens to Fragonard with complete certainty. Over the years, Fragonard's scholarship and skill naturally sharpened, but in some cases there is no way to determine definitively which of the specimens were by Fragonard's hand or his collaborators, such as Hénon. Tragically, Bourgelat erased from the records his taciturn colleague dedicated to building his groundbreaking work in comparative anatomy on the two pillars of perfection in handiwork and the most complete training, but who never sought fame through false pretenses as did so many others.

Among all Fragonard's techniques, it was the dried preparations that led to his notoriety during his lifetime and, thanks to his superb skill, exist more than two centuries later to bear witness to his art. Although the technique of injecting vessels and hollow organs was not new in his era, Fragonard improved on it and executed it so masterfully that each of his works was of exceptional quality. The Italian surgeon Jacopo Berengario da Carpi (c. 1460–1530) was the first to use the technique, in the sixteenth century, going no further, however, than to inject water to reveal the interactions between vessels. It was the seventeenth-century Dutch anatomist Jan Swammerdam who first used wax in place of colored liquids; improving on this technique, Frederik Ruysch used a composition of the finest wax, and he excelled in the art of giving cadavers a plump, graceful, and lifelike appearance. Bernhard Siegfried Albinus (1697–1770), inspired by Ruysch, created magnificent injections,[19] but he kept his formula secret. Fragonard was probably directly inspired by the techniques of Pierre Tarin (c. 1725–1761) and Jean-Joseph Sue *père*, who generally adopted the methods of Alexander Monro.[20] Works that would have been available to him were the *l'Abregé de l'anatomie du corps de l'homme* (A Summary of the Anatomy of Man),[21] published in 1748 by Sue, and *l'Anthropotomie* (Anthropometry)[22] published in 1750 by Tarin or Sue, and its second edition[23] published in 1765 by Sue.

Portrait of Jean-Joseph Sue (père), 1761. Engraving by Le Beau after a drawing by Binet. École nationale supérieure des beaux-arts, Paris.

Prepared specimens were named according to the function of the organ displayed. There were myologies, angiologies, or neurologies depending on whether the dissector-preparator intended to display the muscles, the circulatory system, or the nervous system. There were also splanchnologies, which demonstrated the viscera, and the osteologies, which displayed bones and cartilage.

Fragonard's crucial improvements and perfection of the technique that Jean-Joseph Sue described so precisely in 1765 (see appendix p. 140) have allowed the survival today of a few of the *écorchés* that he himself created. A thorough knowledge of anatomy is evident in these dried specimens. As Fragonard unequivocally declared in his addendum report of July 4, 1792, to the national assembly: "anatomy requires the most exacting work possible only through the most stringent, precise, and thorough training." Rigorously scientific, technically skilled, a perfect master of the discipline, he encouraged the board of public education to take note of "the perfection in quality of our preparations and of our injections." These preparations were the product of time, care, and vast knowledge in achieving each of the numerous complex and delicate steps in their creation.

Portrait of Frederik Ruysch from *Opera omnia anatomico-medico-chirurgica*, Amsterdam, *apud* Janssonio-Waesbergios, 1721. ENVA library

. . .

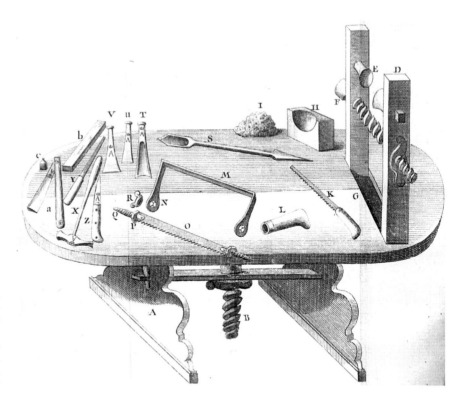

Dissection instruments, Pierre Tarin, *Anthropotomie, ou l'Art de disséquer les muscles, les ligamens, les nerfs et les vaisseaux sanguins du corps humain...* (Anthropotomy, or The Art of Dissecting the Muscles, the Ligaments, the Nerves, and the Blood Vessels of the Human Body...), plate 1. Paris: Briasson, 1750. ENVA library

To create an *écorché*, a body, chosen for its leanness, had its large super-ficial veins cut in several places to drain it of blood, and then it was washed and placed in a heated water bath to warm it in preparation for the injections into the heart and the vessels. The substance injected was a mixture of resin, tallow, oil, and beeswax and was stained red for the arteries, blue for the veins. The German physician Karl Asmund Rudolphi reported that Fragonard's success rate in these delicate injections was six times out of ten, which was considered to be very high.[24] Along with his exceptional experience in dissection and his dexterity at the intricate injection stage, Fragonard's superior work—his "secret," as it were—was owing to the composition of the mixtures he injected into the body cavities. These injections maintained the volume of the specimens, held them in position, and preserved them. Once the body had been injected, it was then dissected as rapidly as possible before decomposition set in. First the body was stripped of its skin, then the dissection was carried out, by means of a scalpel. The dissection procedure varied depending on which organs were to be displayed. After submersion in an alcohol bath, the body or dissected organ was left to dry, supported by a wooden framework, positioned as it would be in its final form. The preparator kept a close watch on the specimen as it dried, rectifying the position of the organs and limbs as necessary, holding them in position by various means: needles, blocks, wedges, and pads of horsehair. The eyes could be either preserved in place and inflated or replaced with porcelain artificial eyes.

Injection of certain organs, such as the vascular system, was completed by corrosion. Corrosion casts were prepared by first injecting the vessels with a corrosion-resistant material, then dissolving the surrounding tissues by submerging the specimen in a corrosive solution. Any remaining tissue was removed by washing or by extremely delicate scraping before the specimen was arranged in its final form. The size and course of even the smallest vessels could be displayed in corrosions.

Because they were primarily interested in veterinary science, the creators of the cabinet at Alfort dissected mostly animals. There was no difficulty finding suitable subjects; horses nearing the end of their lives had no commercial market until horse flesh became a popular comestible in France in the nineteenth century, when old nags came to have slaughter value. However, to create the human anatomy specimens he continually prepared, Fragonard encountered the same supply difficulties as the prosectors in Paris, well described by David Le Breton.[25] In his report of 1792 Fragonard even went as far as to detail a priority in the provision of human bodies for his planned national anatomy museum. "It goes without saying that the principals of the anatomy museum must be authorized to choose in the hospital the subjects that will be most useful to them."

Out of all the thousands of fragile dried specimens produced in the eighteenth century, why is it that only the *écorchés* of Alfort have survived, while all the others have disappeared, ravaged by time or devoured by vermin? In that same 1792 letter Fragonard acknowledged having at his disposal a unique process.

So it is that they [at the school at Alfort] have worked on a great number of preparations in their own possession, which, aside from the advantage of being unequaled in Europe, owing to their unique injection formula, which others would in vain wish to purchase, are blessed with the property of being unassailable by the worms that can destroy all that one has been able to create. And, indeed, I, Fragonard possess some specimens thirty years old that are as beautiful as on the first day they were created.

As well as having mastered the injection phase, Fragonard had improved on finish and protection in his *écorchés*. When dry, the specimens were retouched with colors diluted in varnish: red for arteries, blue for veins, and white for nerves. The finished article was then coated entirely with varnish containing Venice turpentine, a very pure resin extract from the larch tree. Venice turpentine was a precious commodity used by painters to protect their canvases. It is not unthinkable that Fragonard might have obtained his formula from his cousin Jean-Honoré. Perhaps the same varnish was applied to protect both the rictus of Samson and the smile of Diderot (1713–1784), two heroes from the golden age of anatomy—"that science which cannot be learnt from books," as Diderot proclaimed from the platform of his *Encyclopédie*, at the very moment that under Fragonard's scalpel a lifeless body was fashioned to represent Samson.

Fragonard continued to create preparations for the collection right up to his departure from the school in 1771, and his successors carried on this work after he had left. Learned visitors who crowded into the cabinet have left invaluable descriptions of what they found. The great Georg Ludwig Rumpelt, on visiting Alfort in the course of his inquiries into agriculture and breeding in 1779, expressed a mixed opinion. The hostile reception he received from the new director Philibert Chabert did not incline him to be kindly disposed. He praised Fragonard's rigor and honesty and then attributed to these same qualities his dismissal from the school: "[…] it was necessary […] to banish forever from this school all men intelligent, learned, and honest, as was to be the fate of Fraquerac [*sic*]."

If he could not overlook the virtuosity, precision, and refinement of Fragonard's specimens, Rumpelt had scant appreciation for their use to science: "No preparation, even executed with the greatest care, holds to me any value if it does not demonstrate a new fact or the resolution of a debated issue. All their preparations, collected in the cabinet, satisfy only the gaze; few among them serve to demonstrate any new truths, to refute old errors; the refinements throughout are not instructive and demonstrate only the handiwork of their creator."

Rumpelt's bad humor eventually extended to the whole nation, as he went on, sparing no details, to denounce the incurable frivolity of the French: "It is the true character of the Frenchman to let himself be easily carried away by what is new. Hours long he discourses on topics, but at each new idle fancy chases after the first plaything."[26]

And while his compatriot Heinrich Sander also paid compliment to the richness, variety, and orderliness that graced the shelves of the cabinet, he too was astonished by the affectations in the staging of Fragonard's

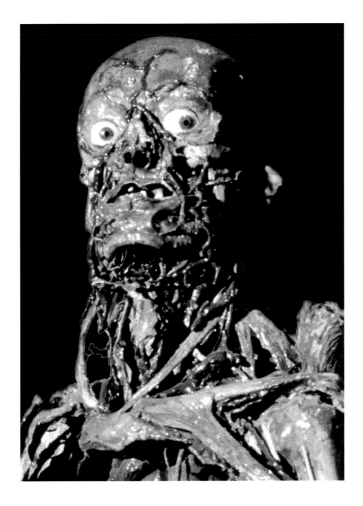

grand subjects: "The frivolous spirit of the nation is manifest, even in undertakings truly grand and noble, where this whimsical and frolicsome character would best be hidden."[27]

Rudolphi, on the other hand, in 1802, proclaimed his admiration: "All the muscles of the animals are intact even as the blood vessels and the nerves too have been preserved. Above all these specimens have made the cabinet famous, and thus I shall examine them most thoroughly."

And both Rudolphi and Sander, as we shall later see, praised the celebrated *Horseman*.

To be sure, the eighteenth century saw the rise of a popular fad in dissection. Public dissections, performed with some of the flourish of theatrical presentation, were available as a daily entertainment. Further, to own an "anatomical cabinet" was considered the very pinnacle of intellectual pursuit and social refinement. This was particularly so among the elite who were caught up in the encyclopédique movement and thrilled by scientific discoveries and technical progress. They loved to collect, among other natural or artificial curiosities, anatomical figures intended to give a lasting insight into the interior complexities of the human body. In his remarkable description of the "most famous cabinets of natural history in Europe," Antoine-Joseph Dézallier d'Argenville (1680–1765) described the great interest of the day in specimens of natural history.

> Cabinet of Mr. Pajot d'Onsembray, of the Academy of Sciences: …& in several long cases one finds skeletons of a man, of a woman & of different animals with several injected specimens.…

> Cabinet of Mr. Geoffroy, of the Royal Academy of Sciences:…several injected human specimens, in particular the head of a young man.…[28]

> Cabinet of Mr. Savalette de Buchelay, farmer-general [customs duties and tax collector]:…the animal kingdom heralded by two dried bodies of a man and a woman, of which all the muscles and the other parts were well displayed; the arteries were painted in red, the veins in blue, and the nerves in white; the primary viscera were shown, as well as the teeth and the nails.[29]

Fragonard's 1792 report to the national assembly is the first extant historical document following a gap of many years after his departure from Alfort. Fragonard surely remained busy in the intervening years and with his unique talent in anatomical preparation was able to profit from the fashion for cabinets of curiosities. No doubt he produced a very large quantity of specimens that were either bought by the wealthy and enlightened members of society or perhaps added to his personal collection numbering some fifteen hundred specimens as of 1792. The exceptional beauty of Fragonard's work garnered him respect and interest not just among high society but among some of the most illustrious scholars of his era. That he was nominated to the prestigious jury des arts by the painter Jacques-Louis David in 1793 marks him at the very least as an eminent contributor in the close and fruitful dialogue between art and anatomy begun so long ago.

Fortified by his vast experience, his fine reputation, and a desire to prove his total allegiance to the institutions of the young republic, Fragonard conceived to crown his life's work with a national anatomy museum. In his eloquent, well-argued report of 1792, Fragonard laid out his rationale for creating a separate institution as a showcase for the science of anatomy, and he asserted the benefits and prestige that such an achievement would confer on both the nation and the institutions that would participate in the initiative. Outlining the pedagogical, scientific, and humanitarian rewards of his scheme, he promised that the cost would be kept to a minimum. Fragonard sketched out a plan both patriotic and republican.

> Even before the Revolution, which has given liberty to France, it was their [the authors of the report] intention to combine their fortunes for the establishment of a museum wherein all the scholars of Europe could access anatomy in all its divisions, man and animal alike, of the highest possible degree of perfection; wherein one could encounter everything needed for study or for reaching even greater discoveries to alleviate the suffering of humanity; wherein the nation's teachers could lead their students to absorb the innermost secrets of the body.

The publication of a proposal for the organization of public education presented by Condorcet to the national assembly in 1792 prompted Fragonard to send to the assembly an addendum to his report, in which he detailed how his own proposal for the national anatomy museum accorded with the legislature's new views on education. In this document, tellingly entitled "Notes essentielles," he took the opportunity to compare the roles and respective merits of the teacher and the researcher. A tinge of the old rivalry between prosector and professor, which long after the Vesalian revolution he had himself endured, echoes in his words. Unquestionably aligning himself with purist practitioners, he emphatically advocated the separation of the functions and argued the equality of status of the academic and the practitioner of anatomy. In a feverish note appended to the text, he avowed the superiority of the practical method of the anatomist in the progress of science over the theoretical speculation of the academic. He reinforced his impassioned proposal with an argument for justice and equality: it is the anatomist, working in fatiguing, extremely dangerous conditions, who pays dearly for his knowledge as he uncovers the facts in his continuous research.

Honoré Fragonard, *Human Digestive System* (stomach and small intestine). Musée Delmas-Orfila-Rouvière, Paris

That the professor is less knowledgeable than the anatomist is a fact in need of no proof. The one cannot do the work of the other just because he has been the other, and while the professor configures his own theoretical physiology based on that which he may or may not have seen, while he reads very eloquently from a text, it is undeniable that his faculties for learning are but stationary. Should it be his good fortune to forget nothing, still he remains at the point of educational advancement where he was [...] when he quit the scalpel.

The anatomist on the other hand, continually hunched over cadavers, advances in his research, realizes his conjectures; each day brings something new, for in anatomy we do not yet know everything; and like practical medicine, it cannot be learned from books.

Fragonard's initiative appears, however, to have stalled at this point without succeeding, and he became immersed in his new position as an expert in the new temporary committee for the arts. Teaching and research in the new institutions had been realigned with educational methods that would improve instruction in the art of healing. Fragonard participated in the selection of specimens from among the collections of the ancien régime to be requisitioned for the new institutions. Surely he would have known all these collections quite well, as they contained many of his own works.

It was in this capacity that he came to examine the cabinet at Alfort and is listed in the catalogues of collections of the Paris School of Health and the National Museum of Natural History. The 1863 inventory of the Musée Orfila (later the Musée Delmas-Orfila-Rouvière) was derived from the collection of the Paris School of Health and lists four specimens from the hand of Fragonard: "A myology of a monkey created in 1797; a stomach and small intestine of man; another small intestine of man; the injected circulatory system of an infant."[30]

Two of these pieces are extant, namely the intestine of a child and the monkey, of the baboon species, in a squatting position, gripping a tree branch in its hands.

As of 1856 the catalogue of the former galleries of comparative anatomy in the National Museum of Natural History listed four pieces originating from the dispersal of the cabinet at Alfort under the Revolution.[31] These are no longer to be seen at the museum. Likely they suffered the common fate of old scientific objects: outdated, lacking relevance to current interests, such things are often disposed of, particularly during a renovation or when staff changes. If the discovery of an obsolete item, decades old, does not stir the emotion or intellect of its finder, its fate, alas, is sealed—into the dustbin it goes, like so much stuff found in the family attic.

To what miracle can the Fragonard Museum attribute its present-day possession of beautiful specimens made by this extraordinary eighteenth-century anatomist and his pupils? For truly it seems miraculous—they bear the marks of rough keeping,[32] and yet today the Fragonard Museum has twenty preparations that have survived more than two hundred years, dating back possibly to the Cabinet du Roi.[33] One can only speculate as to why they were never moved out of the museum.

ANATOMICAL DISSECTIONS AFTER FRAGONARD

Fragonard's work marked the crowning point in the art of dry preparations and, even though they were still produced in large numbers during the nineteenth century, their method was soon to be superseded. Mummification by desiccation, a very old and simple procedure, resulted in a significant loss of volume in preserved anatomical structures, on the order of 25 to 40 percent, depending on the technique used. Even in the eighteenth century, anatomists were beginning to reject this method of body preservation. Félix Vicq d'Azyr noted in his *Instruction* in 1794: "Generally, there are big drawbacks to dry preparations of soft tissues: neither their true color, nor relations, remain. They have lost their form either in shrinking or in excess swelling, and their hardness markedly differs with the natural pliancy of these parts, such that in touching them one learns nothing better than by seeing them. Specimens fixed by corrosion, and the angiologies in general, are the only ones of merit for the anatomist."[34]

Vicq d'Azyr recommended fluid preservation instead; however, this method worked only for small specimens that fit into jars or bottles. For the whole body, the only alternative to dried preservation was the creation of artificial anatomical specimens: waxes or models of plaster or wood, but the material and labor costs were significantly greater. Between the employment of an artist and the expensive raw materials required, the cost of fabricating anatomical models was tenfold more than dried specimens. The scholar's role too profoundly shifted in model making: the anatomist found himself somewhat dispossessed from his work, as the dissection was no longer an end in itself but simply the first stage in the realization of the anatomical model.

However carefully prepared, the body was intrinsically bound to decay, whereas works created by the sculptor, the molder, or the modeler had durability. In handing over his knowledge and technical mastery to the man of art, the anatomist lost his creative role and he knew the dread that any specialist experiences when someone from another trade co-opts his subject.

Of the few preservation techniques available, all were represented in the Cabinet du Roi, so that the student or visitor saw the most complete view possible of the anatomical structure. The collection contained many dried specimens, but the inventory of 1794 lists other preparations as well. For example, number 303 is assigned to a wax figure of a man attributed to Jacques-Fabian Gautier d'Agoty (c. 1716–1785), the French anatomist and engraver known above all for his sublime large anatomical color plates reproducing dissections in all their natural splendor. Along with the dried specimens, the collection also contained a great many bottles with fluid-preserved specimens. There were also taxidermic animals and "monsters" preserved as wet specimens or dried, although the mummification technique was eventually abandoned in the nineteenth century, especially among collections open to the general public.

Plaster models, some used in the study of farriery, others for determining the age of a horse, and still others demonstrating anatomical structures, were also in the collection. Some of the plaster or wax models could better serve the beginner than hardened flesh, although, on the other hand, desiccation was still used to preserve examples of particularly fine quality dissections thought to be worth preserving for the sake of demonstrating the dissection process itself, or for study by specialists in anatomy able to look past the distortions and the altered appearance of the organs.

All of these specimens had come in the twentieth century to be known as "pièces de cabinet"—an indication of the purpose they served. Their production remained very high throughout the nineteenth century among the faculties of medicine, leading the Musée Delmas-Orfila-Rouvière in Paris and the Conservatoire of Anatomy in Montpellier to own an impressive number of nineteenth-century specimens. Often created in recruitment competitions for new teachers or prosectors, these specimens are examples of dissection serving the prestige of the institution and its instructors rather than didactics.

Over time, the collection of the cabinet at the school at Alfort reflected the decline of the desiccation technique of preservation. In 1902, at the opening of the third museum at Alfort, the main collection consisted mostly of skeletons, wet-preserved specimens in jars and bottles, and models made by Eugène Petitcolin (1855–1928). The discovery in 1867 by August Wilhelm von Hofmann (1818–1892) of a molecule with almost miraculous properties of fixation resulted in the

Gaetano Giulio Zumbo (1656–1701), *Human Head*. Wax. Musée Delmas-Orfila-Rouvière, Paris

Corrosion cast of the arterial sytem of a dog. Department of Anatomy of the Veterinary Faculty of the University of Utrecht

compound formalin, which had greatly aided the practice of dissection in embalming by slowing decomposition, improving hygiene, and reducing the nauseating stench.

TRANSFORMATION AND POSTERITY

Although the historical records up until the eulogy by Michel-Augustin Thouret, director of the Paris School of Health, are puzzlingly devoid of information about Fragonard, in his lifetime his work earned him renown among the elite, scholars, politicians, artists, and the echelons of high society, even while it did not spare him cold and ruthless maneuvering from Bourgelat. In praise of Fragonard's simple morals, Thouret's anecdote about Fragonard's "poor" attire nonetheless betrayed Fragonard's lack of social refinement; Thouret also suggested that in some way his work in the Paris School of Health had become less than satisfactory, owing to his exhaustion and life-changing reversals of fortune. Fragonard's enduring devotion to anatomy, though revered throughout the scientific community, was scholarly and yet silent—"he wrote nothing," Thouret stated, and became increasingly silent and "less communicative," and yet, "with imagination he created new specimens" and presented them for the edification and admiration of his visitors. It is conceivable that Fragonard's silence grew in direct correlation to human anatomy coming to have much less to be said about it now that so much knowledge had been gained in a discipline that depended on descriptions of the macroscopic form.

During the nineteenth century the institutions that housed the anatomical collections Fragonard had created, enriched, and valued all his life and had hoped to use to establish a national museum began to lose pride in these objects. The cabinet at Alfort was expanded after the Revolution, and its mission changed. In 1828 it became a "cabinet of collections," a repository of objects that anatomy instructors could use in demonstrations to their students in the highest form of nineteenth-century university education: the lecture. In this capacity the collection assumed such importance that twice it was moved before being re-formed into the museum that thrives yet today. The collection of the Paris School of Health became the first Musée Orfila, then the Musée Orfila-Rouvière, then the Musée Delmas-Orfila-Rouvière, in accord with the patrons who assumed its management and accessioned into the museum their own collections. In the second half of the twentieth century, like nearly all scientific work of the past, it came to be regarded as outmoded and obsolete. Technological advances resulting in computer-generated 3-D imaging and multimedia in the 1990s made for a sudden and swift turn away from some standard methods of teaching, and the old collections seemed terribly dated by comparison. A good many anatomists and surgeons saw a new El Dorado where avoiding contact with the dead body would eliminate all ethical and hygienic challenges intrinsic to cadaveric dissection.

Learning center of the Department of Anatomy of the Veterinary Faculty of the University of Utrecht

Some two decades after the technological revolution the results are mixed. The limitations inherent in these new teaching methods have become apparent, leaving a large segment of the scientific community in a quandary. The educational value of such old scientific collections today is acknowledged although it has shifted. The methods of preservation from the past are obsolete, and yet they are historically relevant to their successor, the technique of plastination, from which a flourishing establishment of extensive new teaching collections has arisen. Cabinets overflowing with plastinate-preserved organs today have a place in both practical classes and self-directed study. Plastination allows the student to handle a genuine, completely safe anatomical specimen, its volume and proportions intact. Renovations in universities and schools today almost routinely include a learning center, an open space where students can work, connect to databases, access books, and, in the case of anatomy, handle classified and annotated specimens just as if they were books. This educational approach, modern in practice, nonetheless still bears an unmistakable affiliation with the cabinets of the eighteenth century.

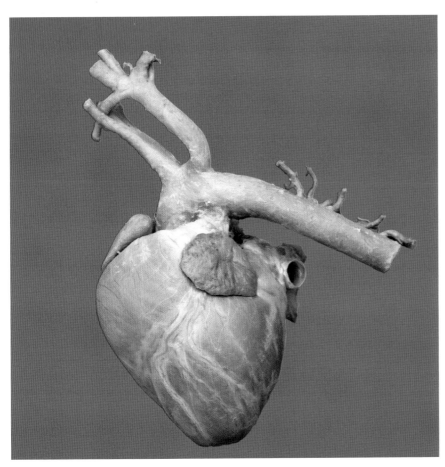

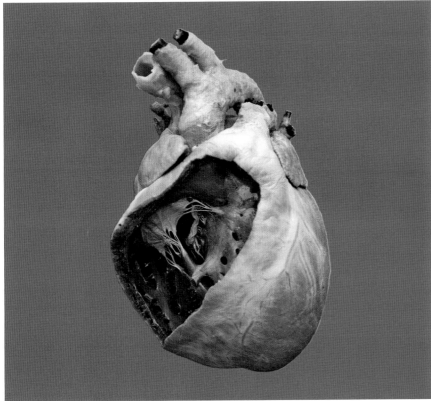

TOP: Left atrial surface of a dog's heart. Plastinated organ. ENVA

BOTTOM: Dog's heart in which the windowed right ventricle reveals the right atrioventricular valve. Plastinated organ. ENVA

The *Écorchés*

These *écorchés* were described in 1794 by Fragonard himself in his inventory of the cabinet at Alfort. Each entry begins with a description of whatever he found particularly striking. The original manuscript is in the National Archives in Paris.

Each specimen was equally well described by Karl Asmund Rudolphi when he visited the collection in 1802. Rudolphi indicated that the majority of the items, and in particular the largest, had been preserved by Fragonard,[1] who was "so skilled in the art of injections that six out of ten were carried out with success"—quite a feat given the difficulties of injecting cold cadavers with molten substances.

Some among the twenty-one *écorchés* remaining at the museum today are known to have been produced by two of Fragonard's pupils who succeeded him: Jacques-Marie Hénon and Pierre Flandrin.

See p. 153.

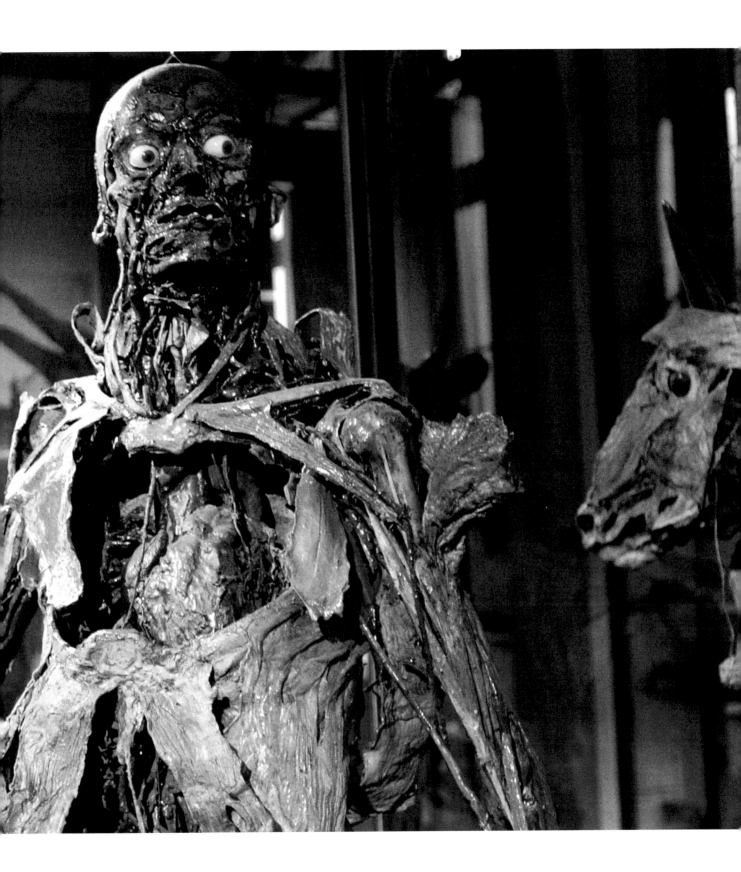

The Écorchés *Preserved at the Fragonard Museum*

The Horseman
14th case, enclosed all four sides
 " (horse) 1322 entire horse mounted
by a myology of a man, catalogued under the
same no.
1. all the muscles in general
2. the principal nerves that traverse all
   parts,
3. the heart, the aorta,
4. the arteries and pulmonary veins,
5. the carotid arteries and the jugular
   veins,
6. the esophagus, the tracheal tube,
7. the penis and its venous network, the
   nerve,
8. the seminal vesicles, the vasa deferentia,
9. the arteries, veins, and nerves of the
   spermatic cord, the testicles,
10. the bladder

The *Horseman* is certainly the most famous specimen in the museum, and it is incontestably by the hand of Fragonard. It has frequently been the object of interpretations in all probability groundless, and often been called the "Horseman of the Apocalypse"—in explicit reference to the work of Albrecht Dürer (1471–1528).

Rudolphi described it in 1805 among the dried preparations[2] as the jewel in the cabinet's collections:

"a. dried preparations

"All the muscles of the animals are intact even as the blood vessels and the nerves have been preserved.

"Above all these specimens have made the cabinet famous, and thus I shall examine them most thoroughly.

"a.a. a large horse with the muscles of generally good appearance, and at least not been eaten by maggots. The arteries are colored red, the veins blue; the trachea white, the esophagus brown;

the heart has been preserved, as well as the pulmonary vessels and the largest vessels of the mesentery (…) the attitude of the horse is as in a gallop and it carries on its back an *écorché* holding in his hand reins of blue ribbon."[3]

The objective of this dried specimen is to display the myology, that is, the muscles of the horse and the man, but the notes listing the structures intended for study in this work clearly refer to the horse. The props vanished at some point after the 1950s, but the staging is simple: the horse rears up as in a semigallop, and the man gazes straight ahead into the distance; in his left hand he clutched the blue silk reins while in his right hand he lifted a whip,[4] a most striking scene. Heinrich Sander also described the *écorché* as having been surrounded by "skeletons of horses [that] carried horsemen—small skeletons of boys holding reins of blue silk passing through the horse's jaw, and a whip in hand. The frivolous spirit of the nation is manifest, even in undertakings truly grand and noble, where this whimsical and frolicsome character would best be hidden."

These little boys were in reality foetal skeletons listed in the 1794 inventory under the numbers 1402 and 1404:

Human 1402 myology of a human foetus on horseback, on the foetus of a donkey, holding reins in hand

Horse 1404 myology of a human foetus, mounted on a foetus of a horse, holding a whip in hand

One can picture the scene of this great horseman surrounded by his tiny squires eternally riding in death alongside such

astonishing subjects as the complete myology of a reindeer,[5] a stag with six horns, a human foetus holding a heart in its hand,[6] and human skins.[7]

One legend about the *Horseman* has persisted. Rudolphi recounted that, according to an account of a journey published in Germany in the eighteenth century, the *Horseman* was in fact a horse *maiden*, a young girl whom Fragonard loved but whose unyielding parents opposed her union with Fragonard and she died of grief. The story has it that Fragonard had dug her up, flayed her, dissected her, and preserved her. In support of the story it is said that when a foreigner visiting Alfort questioned him, Fragonard's only response was a look of profound melancholy—the anatomist had given his true love the ultimate gift: eternal preservation.

Fortunately Rudolphi was able to question Godine, the librarian at Alfort at that time, who would know how the story got started. Disturbing rumors about the death of a grocer's daughter had begun to circulate just after her burial, leading to her exhumation. Very shortly afterward, the *Horseman* appeared in the cabinet, and the rumor spread among the general public that it was the grocer's daughter—as if so elaborate a preparation could be made in a few days!

As was noted by Rudolphi, the *Horseman* prepared by Fragonard is indeed a boy, as is evident from the remains of the penis that Fragonard removed better to seat the young man on his mount. The legend has, however, traversed the centuries; numerous journalistic articles have peddled this outlandish story.[8]

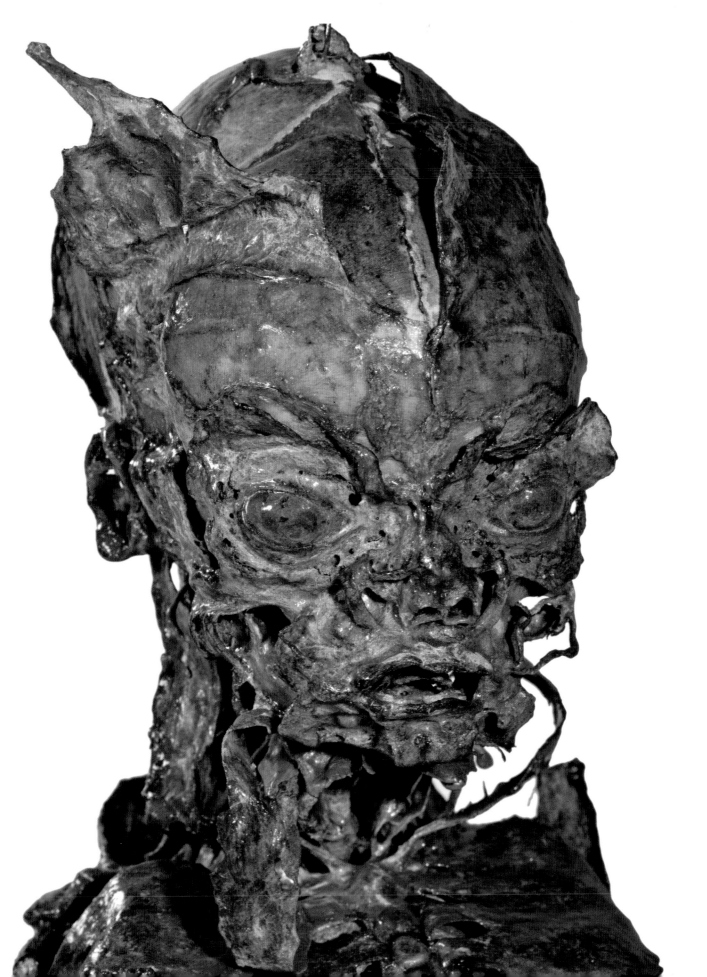

The Foetuses

Because several similar objects are described, these pieces cannot be positively linked with the 1794 inventory descriptions.

The Fragonard Museum collection includes four human foetuses. The first is a dancing foetus: suspended by a thread, it throws its right leg up in the air while the left leg is bent beneath it, as if dancing the "Gigue," a baroque dance taking its name from the English "jig." This danse macabre may seem shocking to the viewer. To the side of this *écorché*, a group of three foetuses walk together on a wooden base. Their vascular systems are injected with red and blue wax and their craniums are open to reveal the vessels beneath the fontanelles, a swirl in the cranial cavity. The anatomist has cleverly re-created these vascular tufts with metallic wire coated in colored wax. The impression is tragicomic: the figure on the right advances, chest slightly bowed, arms akimbo; the figure on the left, approximately the same height as the one on the right, is secured from behind, the vertebral column in extension and the abdomen open wide, revealing the vascular network of the liver.

Most surprising of the three is the central figure, smaller than its fellows, appearing to be running bandy-legged at a slow trot.

These were not the only foetuses in the Cabinet du Roi. There is a description, for example, in the inventory of 1794 of a foetus holding its own heart in its hand, in the tradition of the seventeenth-century anatomists who, following the work of Frederik Ruysch, arranged the little bodies of these babies together in a macabre tableau.

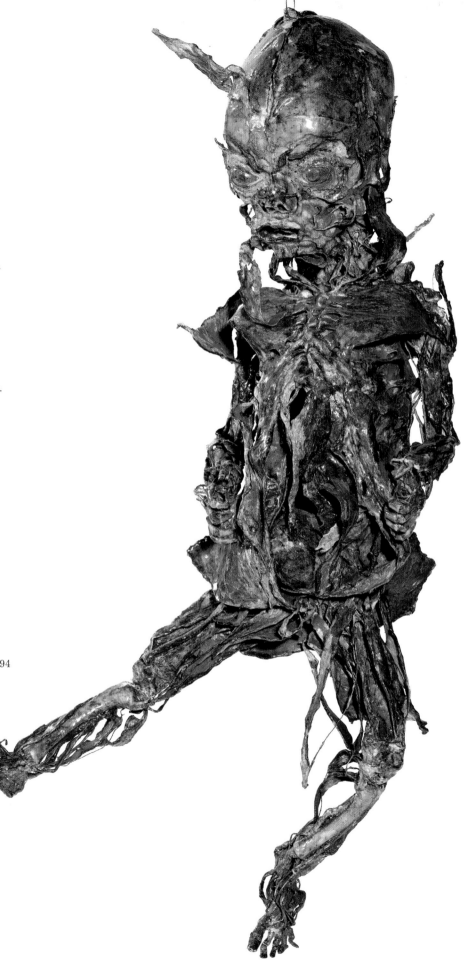

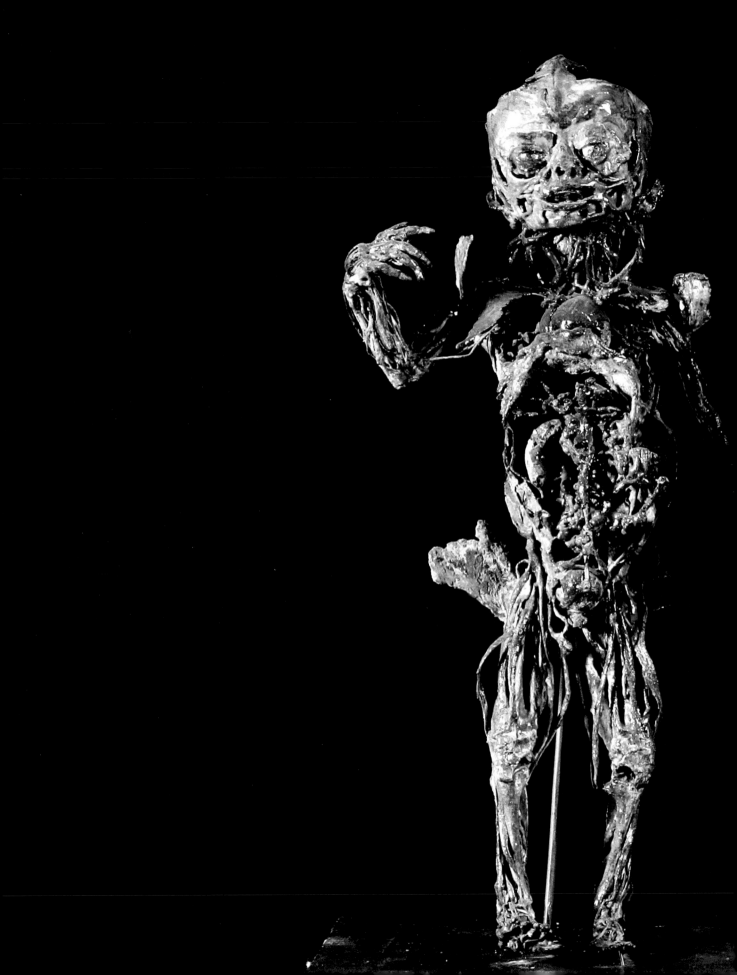

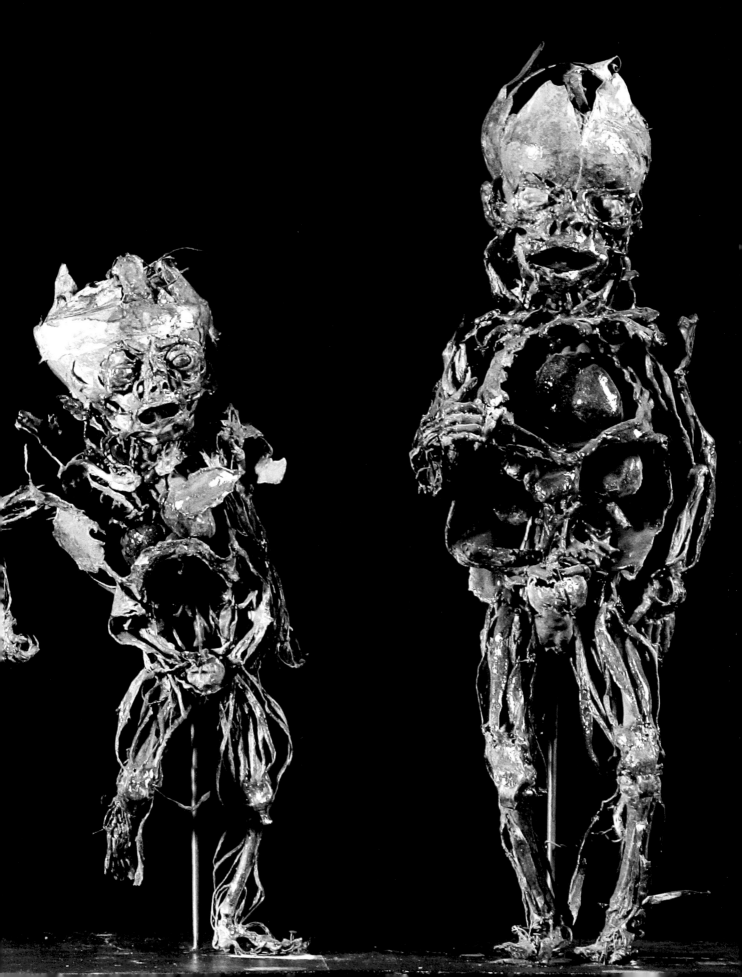

Man with a Mandible

Human   1149   myology and angiology of a man holding a jawbone in his hand

1. the heart, and the aorta
2. the arteries and pulmonary veins
3. the carotid arteries
4. the jugulars, which spread over the face
5. the vena cava, the branches of the hepatic (the fifth missing)
6. the mesenteries
7. the kidneys, the bladder, the ureters
8. the reproductive organs
9. the veins of the upper and lower extremities
10. the principal nerves of these same extremities; beautiful specimen.

In technique, this *écorché* is probably the most dazzling in the collection. Before the Revolution, this man was probably displayed in the amphitheater adjoining the cabinet, behind protection of a glass case. Sander recounted in 1783 that "the amphitheater adjacent is more beautiful, certainly, than many an 'école ducale' in Germany, where one gets covered in dust and sometimes risks one's life. Benches line the walls of the amphitheater. Standing tall in a glass case is an injected human body, probably intended to demonstrate the fundamentals of human anatomy before one passes on to comparative anatomy."[9] This *écorché* is certainly different from the *Horseman* that it faced. While the *Horseman* rides in peace, the *Man with a Mandible* is violent, savage. It is the most beautiful example of sculpture using the anatomy of an actual body to express emotion. In this evocation of Samson battling the Philistines with the jawbone of an ass,[10] Fragonard sought to convey all Samson's aggression and rage. The body stands erect, intent on dealing a death blow. His porcelain eyes look sidelong, gauging the enemy; his lips are twisted with the rictus of effort. Bent backward behind him, his right arm readies to strike. His aggression is heightened by his exposed viscera, injected with bright colors. His vermilion heart gleams from the center of the arterial and venous branches of the pulmonary system. The mesenteric arteries form clusters in the abdomen, while the penis, stiffened by wax injection, too heightens the violence of the figure. The man is clearly aged: only a few teeth remain, and a total absence of fat (best for creating the finest specimen) suggests he died in a state of advanced emaciation. What ailment might have assailed him? We'll never know.

The Human Busts

Two human busts remain in the collection, each very different from the other. The first is described in the 1794 inventory under the number 1161.

Human   1161   human bust with:
1. the muscles
2. the carotid arteries and their distribution
3. The jugular veins and their distribution on the face
4. the salivary duct
5. the aortic trunk, the azygous vein
6. the brachial and intercostal nerves

This specimen is distinct from others of its type in that it does not display the brain's vascular system. The thorax is wide open and clearly presents the aorta and the superior vena cava, injected with red and blue wax. Fragonard has preserved the natural eyes so well that their gaze, turned upward, is less unsettling than that of the man with the mandible. Half open, the mouth reveals a set of fine teeth. Injected veins cover the head, forming a network that spreads across the temples, face, and forehead. The cranium is perforated by two simple holes that probably served for extraction of the cerebral matter, necessitated by its putrefaction. This bust seems finally at peace.

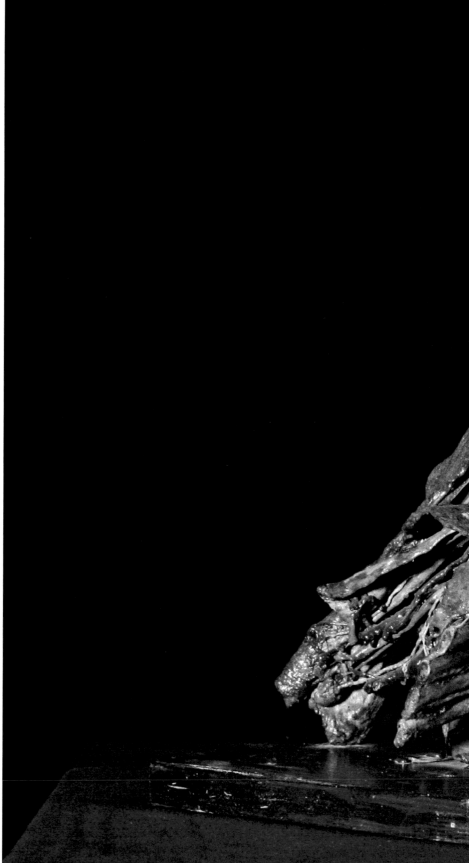

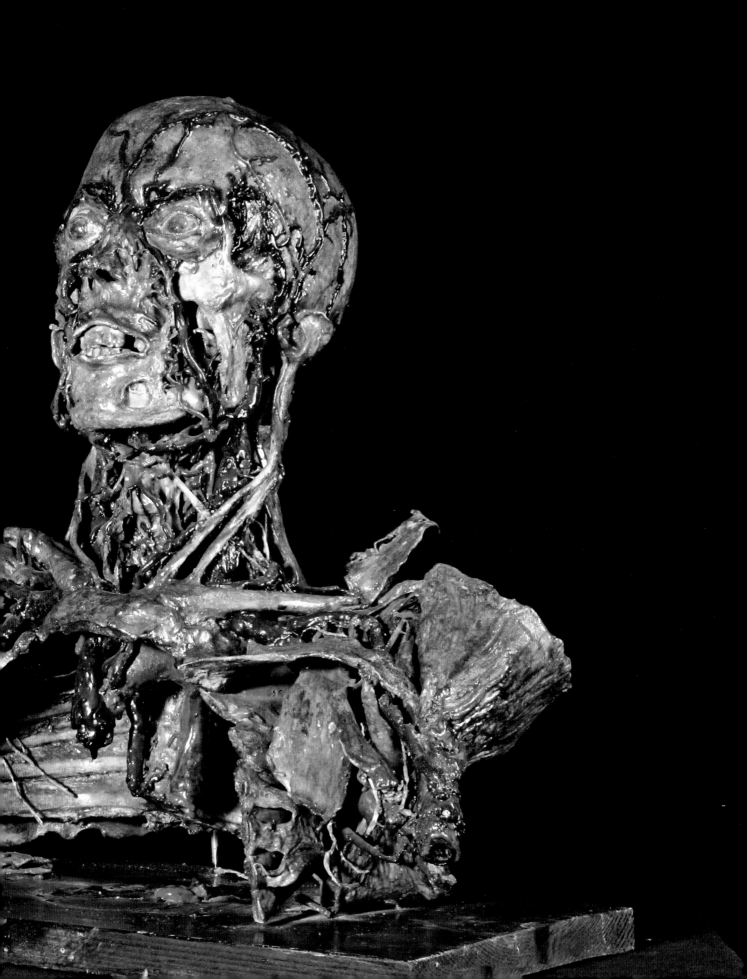

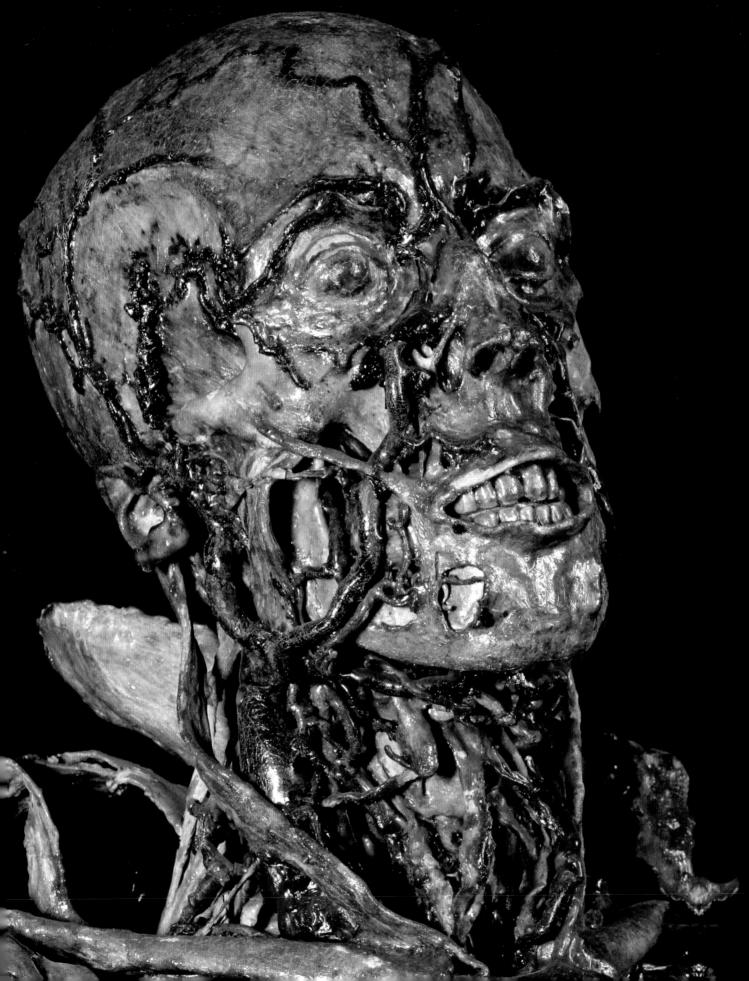

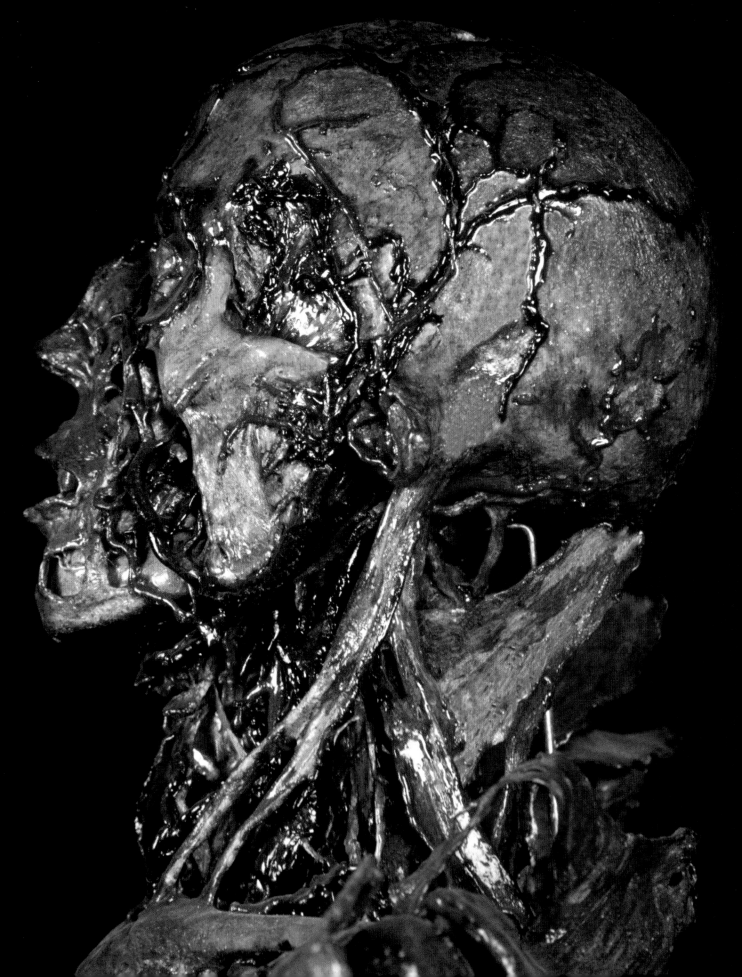

Human 1163 human bust, myology, displays

1. the muscles
2. the arteries and the veins distributed to the head
3. the cranium is open, the sinuses of the dura mater and the vessels of the brain visible,
4. the principal nerves of the face and the neck,
5. the two Stensen's salivary ducts

In contrast to the previous bust, this bust, probably number 1163[11] in the 1794 inventory, seems rather tormented. The jaw appears clenched in a death spasm. Eyelashes surround the sunken orbicular muscles of the left eye and the membranes covering the cranium fold over the front, primarily on the left side. The cranial cavity has been opened with a saw, preserving an arch that attaches the falx cerebri, still intact and supporting the venous sinuses of the dura mater. The anatomist made two attempts before succeeding: the trace of a first saw cut is visible on the cranium before making the final cut. These two bust specimens perfectly demonstrate the anatomy of the region.

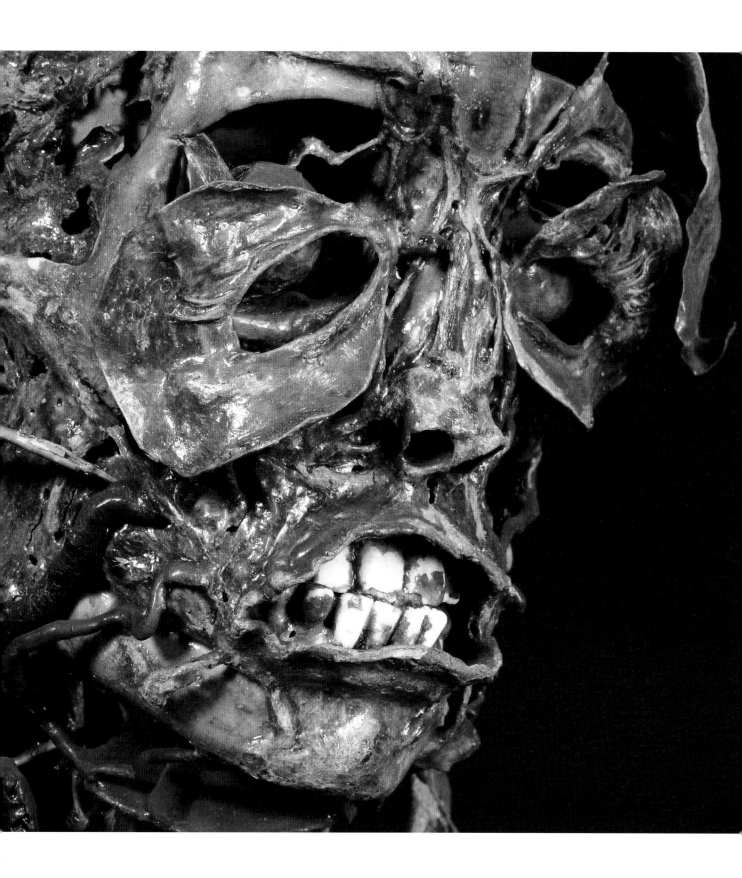

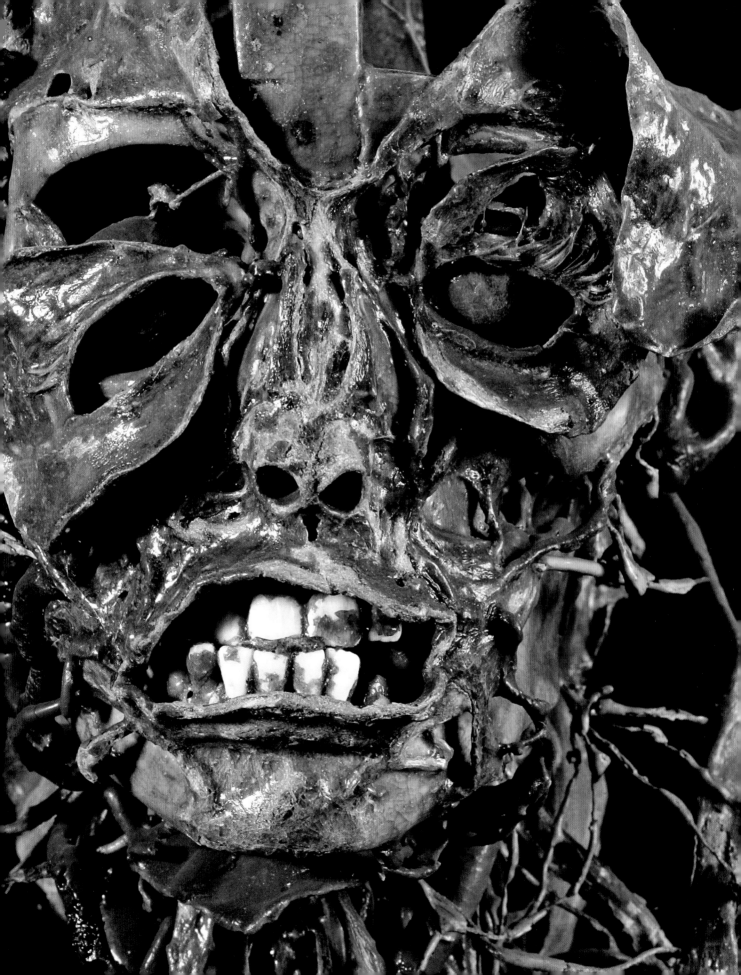

The Human Limbs

Because several similar objects are described in the 1794 inventory, the human arm cannot be positively identified.

OPPOSITE

Human   1188   inferior extremity of man with the pelvis, visible are:

1. the muscles
2. the nerves
3. the arteries
4. the veins which fully envelop the extremity down to the toes; a superb preparation in which numerous anastomoses of the veins are visible.

These specimens of a human leg and arm demonstrate the muscles as well as the vessels and nerves that traverse them. These dissections typify Fragonard's dehydration technique, which resulted in shrinkage of the organs and a parchmentlike appearance of the limbs.

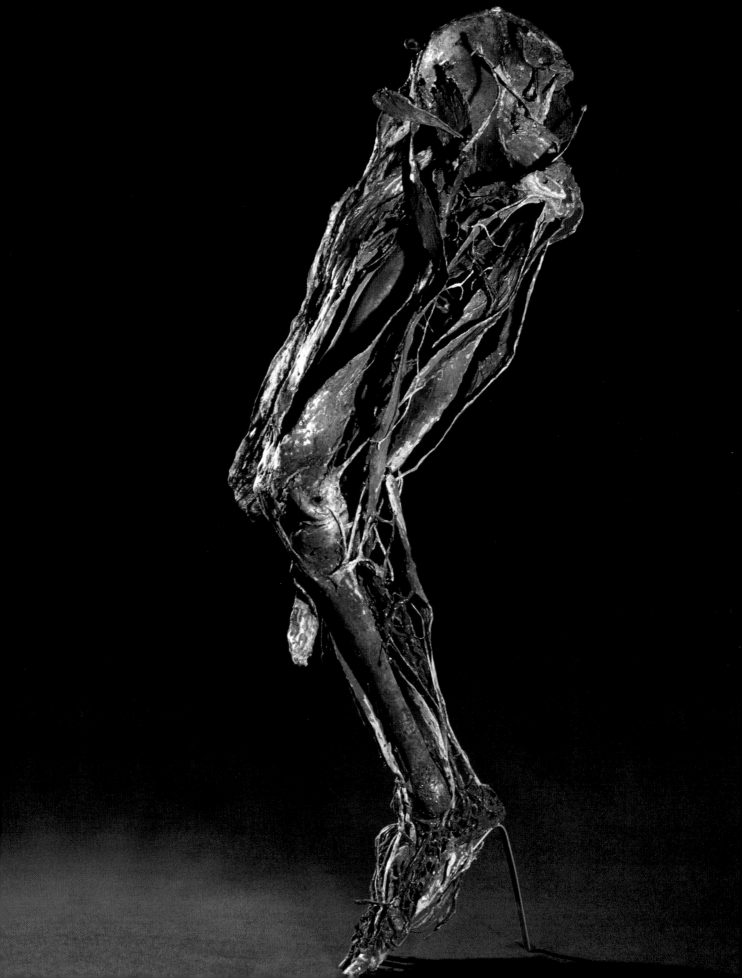

The Corrosion Casts

RIGHT

angiology   1007   human angiology, visible are:

1. the heart and the aorta leaving it
2. the carotids and their distribution to the head
3. the axillaries and their distribution to the superior extremities
4. the mesenteries, the intercostals
5. the ureters, the bladder with its arteries
6. the ilia and their distribution to the inferior extremities,
    the whole a fine specimen mounted on a board

OPPOSITE (DETAIL)

Female   1006   arteriotomy of a woman, the following parts visible:

1. the duct of the gallbladder, the cystic duct, and the bile duct
2. a portion of the duodenum
3. the pancreas
4. the pancreatic duct, which opens into the intestine
5. the spleen
6. the kidneys, the ureters
7. the bladder
8. the womb, the uterine tubes
9. the ovaries giving rise to the round ligaments
10. the broad ligaments
11. the rectum, the pubic bone
12. the external reproductive organs
13. the hymen
    the whole beautifully presented on a board

These specimens were intended solely for study of the circulatory system. They were created from an internal cast of the arteries in the body of a child and of an adult. It appears that these common-place specimens were prepared at the same time: the simple boards they are affixed to are covered in wallpaper.

The technique is not described in the records, but Fragonard probably injected the vessels with wax, which would have preserved some pliability. The soft tissues were then corroded and the vascular system positioned flat on the boards.

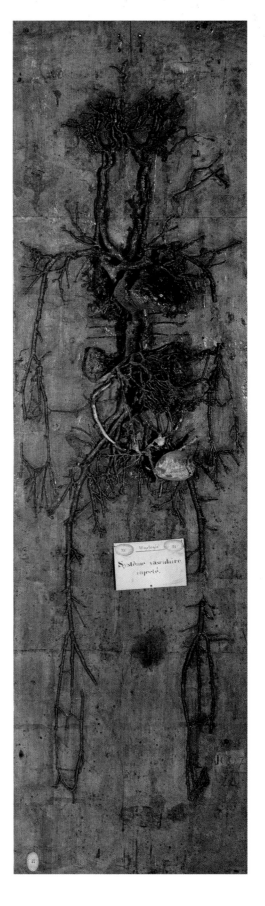

Nilgai. From Georges Leclerc de Buffon, *Histoire naturelle* (Natural History), supplement to vol. 6, plate 10, p. 114. Paris: Imprimerie royale, 1782. ENVA library

The Nilgai

Doe of the Indies   1448   myology of the doe of the Indies, visible are:

1. the heart, the aorta, the pulmonary arteries and veins,
2. the carotid arteries and their distribution to the head,
3. the jugulars and their divisions at the head,
4. the tracheal tube, the esophagus
5. the veins and arteries of the anterior and posterior extremities
   Beautiful piece

Traveling in 1773 to England to obtain horses, Flandrin wrote of, among other animals, two llamas, and in a letter of October 17, 1773, to Bourgelat,[12] of "a stag with the horns of an ox from the Indies," the nilgai.[13] "I do not know if we would need a permit for everything coming to us from London. We can bring, at least I believe, along with twelve stallions, a ram and a sheep, a billy goat and a doe, a bull, a cow, and a stag with the horns of an ox from the Indies, as well as several geese, ducks, many, many cocks and hens, doves, peacocks, and pigeons. With warmest wishes and best regards." Posted at the top left of this letter with the date October 19, 1773, is "permit to Bourgelat."

The nilgai is especially interesting for its preparation technique. The cutaneous trunk muscle has been dissected and perfectly preserved, spread like a blanket covering the animal. The projecting jugular vein is filled with brown wax and painted with blue wax. The open thorax reveals the heart painted brown, its apex sutured with horsehair, thus showing the means by which the anatomist made the injection to fill the vascular system.

A thoracotomy has permitted dissection of the fine branches of the bronchi, forming a complex network.

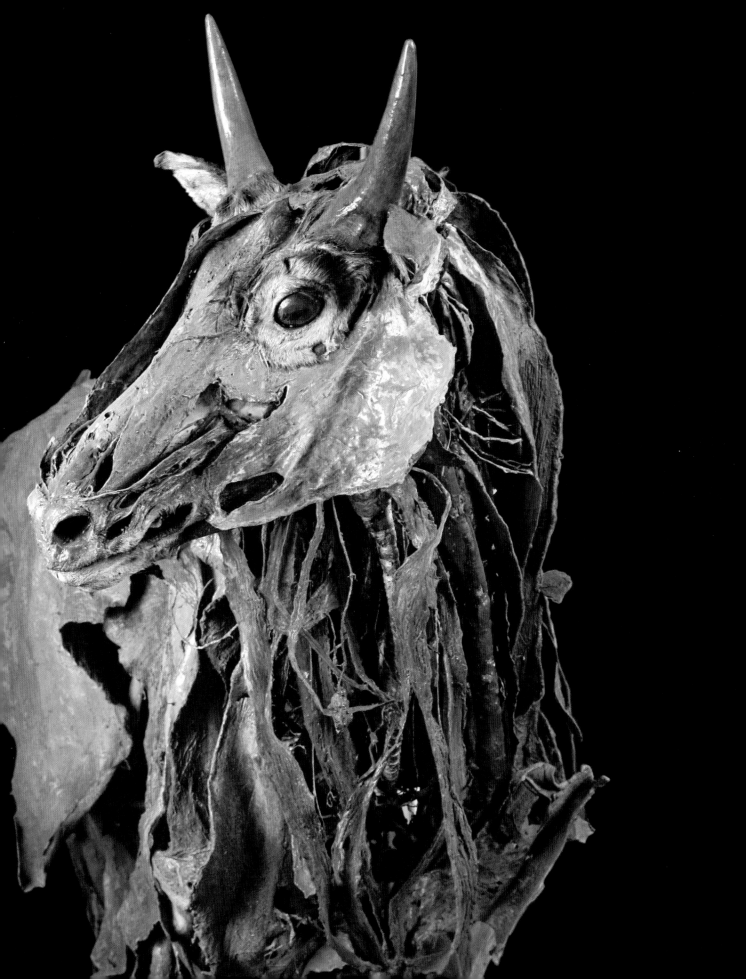

Llama. From Georges Leclerc de Buffon, H*istoire naturelle* (Natural History), supplement to vol. 6, plate 27, p. 206. Paris: Imprimerie royale, 1782. This llama must be the young of the dissected female preserved at the Fragonard Museum.

The Llama

Flandrin reported on seeing a female llama and her young in his travels in 1773.

Llama or alpaca   1501   the myology of the llama or alpaca female, visible are:
  1. the heart
  2. the arteries and pulmonary veins
  3. the aorta and vena cava
  4. the carotid arteries and jugular veins and their distribution to the head
  5. the esophagus and tracheal tube
  6. the arteries and axillary veins, which run as far as the hooves
  7. the crural arteries and veins, which run as far as the hoof of the posterior extremity
  8. the principal nerves of the head of the neck of the extremities
  Beautiful specimen

The great naturalist Georges Leclerc de Buffon first described the llama in volume 13 of his *Natural History*, published in 1765,[14] in an article that is a bibliographic synthesis culled from old texts and lacking illustrations, Buffon having never seen a llama. In the 1782 supplement to volume 6 of his *Encyclopédie*, however, is a new article composed after Buffon was able to study a specimen that arrived from England with its mother in 1773. Buffon supplied the key to the enigma: "shown here (plate 27), the figure of a llama, drawn from life, and still living (August 1777) at the veterinary school at the house of Alfort. This animal was brought from the Spanish Indies to England and was sent to us in November 1773. It was still young, and its mother accompanied it but died shortly after arrival; its taxidermic skin and its injected body stripped of skin can be seen in the beautiful anatomical cabinet of Mr. Bourgelat."[15]

The female llama on display today arrived at Alfort two years after Fragonard's departure and was prepared by his successor Hénon. The muscles, thin as parchment, give the *écorché* a delicate appearance. The vessels are injected and the nerves clearly visible. The trachea, broken, shows the horsehair used to stuff it before dehydration.

OVERLEAF

The Bust of a Sheep with Four Horns
Sheep   1162   myology and angiology of a sheep, visible are
  1. four horns
  2. the heart and the principal vessels leaving it,
  3. a network of veins at the tip of the nose

The museum contains several specimens of Jacob's sheep, a breed developed in England in the eighteenth century. Its white fleece spotted with black and its horns made it desirable as an ornamental animal in the gardens of the wealthy English. This bust displays a particularly fine-quality dissection with vessels injected, the heart dissected, and the bronchi cleaned. The specimen is displayed suspended in the air so that one can see through the thorax to view the anatomical structures inside it: the heart, the great vessels, and the intricate architecture of the broncho-pulmonary tree are still intact.

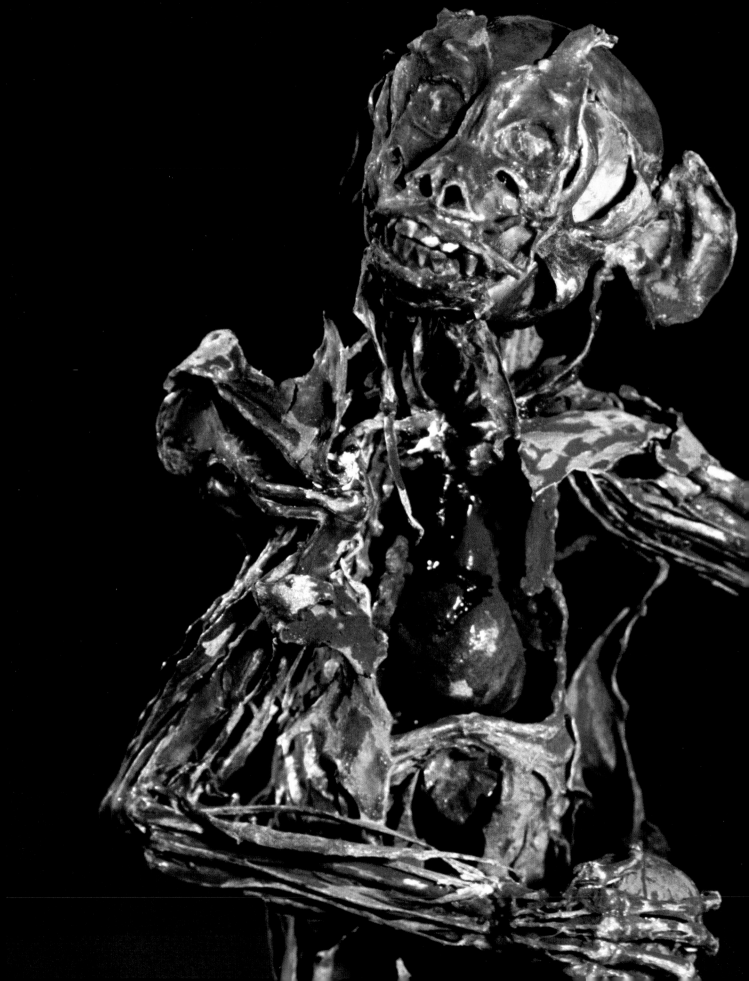

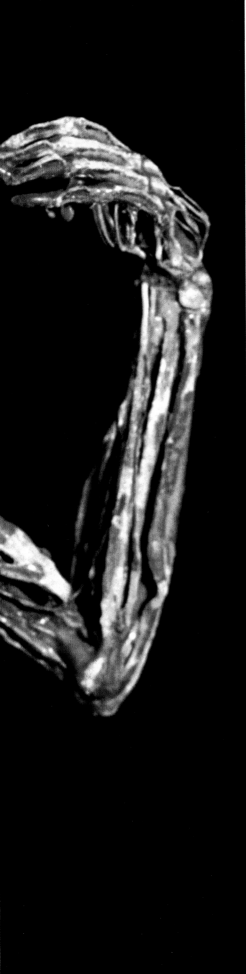

LEFT AND PAGE 117

The Monkey Holding a Nut
Monkey   1148   myology of a monkey
holding a nut in its hand

Little green monkeys such as these were
frequently seen in eighteenth-century
markets.

   This one was precisely described by
Fragonard in 1794: in its right hand it
holds a nut while its left hand is raised
and its face turned skyward. As with the
human *écorchés*, its thorax is wide open,
and its heart, injected and painted, is
immediately visible. Fragonard used the
same techniques for these animals that
he used for human specimens.

PAGE 116

The Dancing Monkey
Green monkey   1151   myology and
angiology of green monkey showing the nerves
of the superior and inferior extremities

This second monkey is posed deferen-
tially. Its hands are joined as in applause,
its lips are drawn back in a tragic grin,
and its large ears are attentively turned
frontward. The opening in the cranium
is evidence of the means of removal of
the brain.

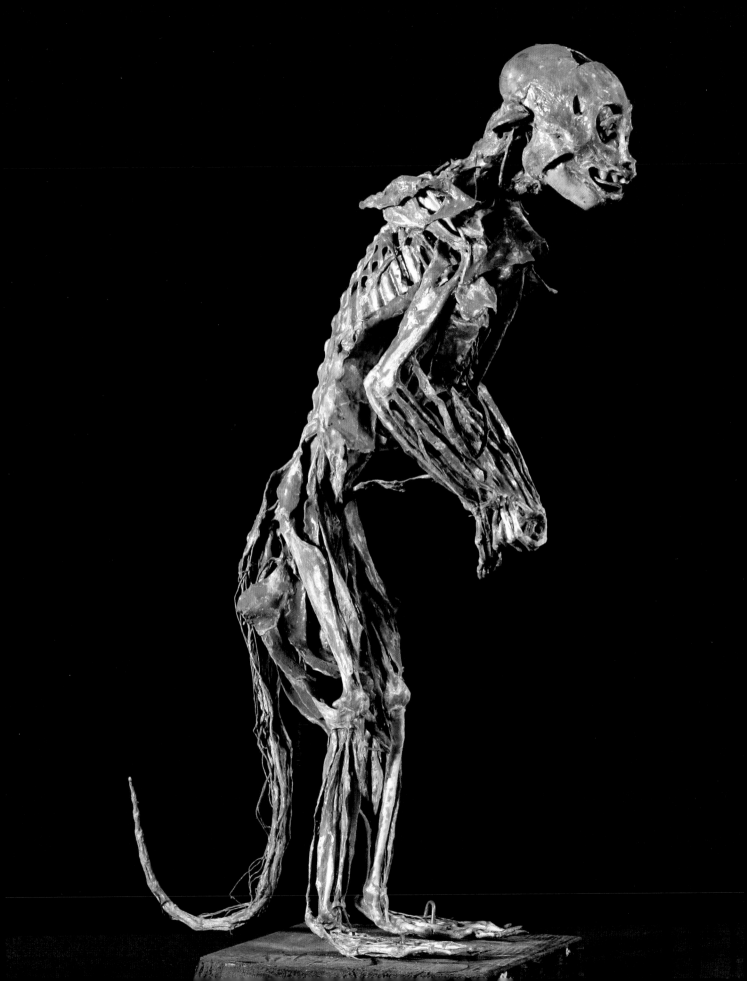

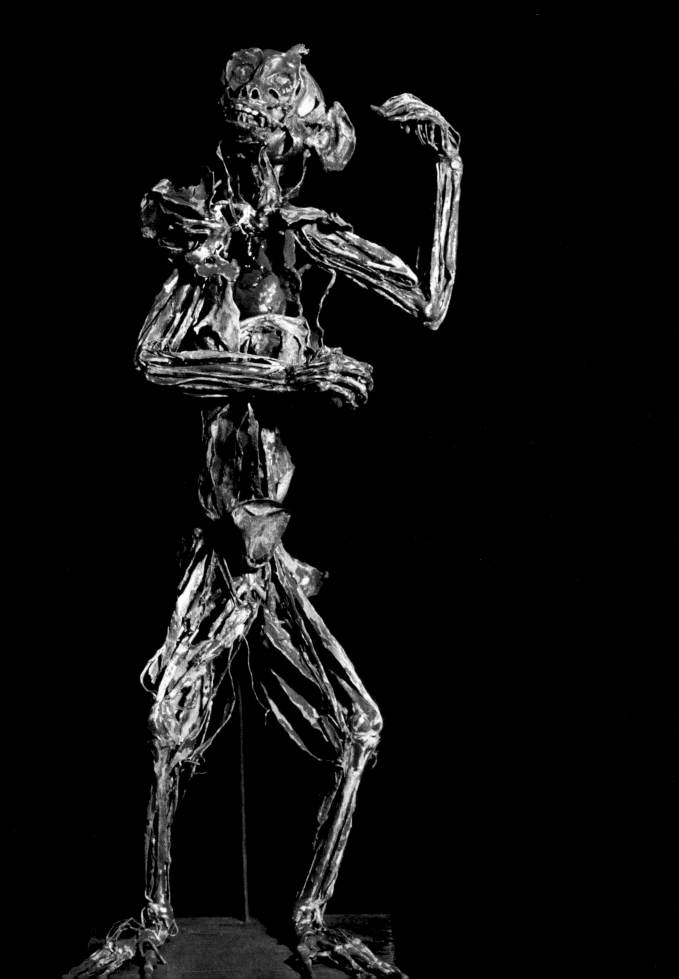

écorché de marsouin par Fragonard

The Porpoises

Two *écorchés* of porpoises are in the museum. Even though they were dissected and preserved by the classical method for dry preparations, they are probably not Fragonard's own work, but rather were likely made after a trip Chabert and his pupils made to the seaside in the 1780s, from which they returned with a number of fish and marine mammal specimens. The bodies and organs of these creatures duly took their place in the cabinet at Alfort, enhancing it as a cabinet of comparative anatomy and natural history.

Of the two porpoises, most beautiful is a female that welcomes visitors at the museum's entry, displaying a finely dissected vascular system.

The Neurologies

These are taken from dissections of the nervous system of the horse, or rather from the conservation of the meninges, the fibrous membrane that surrounds the nerve and gives form to this very fragile tissue.

The first specimen (above) is the brain of a horse in which all the cranial nerves up to their terminal divisions are preserved. This brain could certainly not have been simply extracted from the cranium; the whole organ must have been dissected out, the superficial nerves first isolated by dissection of the muscular layers, then the deep nerves freed by scraping the bones of the head, breaking them up, and removing them piece by piece to reveal the finer branches trapped within the massive cranium.

The second specimen (not pictured here) is of the same type; it is taken from the whole nervous system, central as well as peripheral, which has been freed from the trunk and limbs. The anatomist has preserved everything: the brain, the cranial nerves, the spinal cord, the spinal nerves up to their extremities, and even the fibers of the autonomic nervous system, notably the sympathetic ganglia.

Their considerable dissection work gives these specimens great value. The cranium and vertebrae have been morseled, the muscles dissected to extract the segmental nerves; the bones of the limbs around which the nerves wind have been extracted; the ganglia between the intervertebral muscles have been uncovered—all this painstaking freeing of the nerves and of the spinal cord must have been done manually with razor-sharp instruments that could just have easily damaged the nervous system, at so much greater risk of destruction than bone. What's more, it must all have been done very rapidly because putrefaction was difficult to control; formalin was as yet unknown and the tissues to be conserved were merely soaked in alcohol. The preparators of this specimen would have had to work in the cold of winter to limit tissue decay. The speed and extreme precision required to create these specimens would have been a challenge to the anatomists—inspiring the rapturous response of Louis Petit de Bachaumont (1690–1771) upon seeing this specimen, which he credited to Pierre Flandrin.[16]

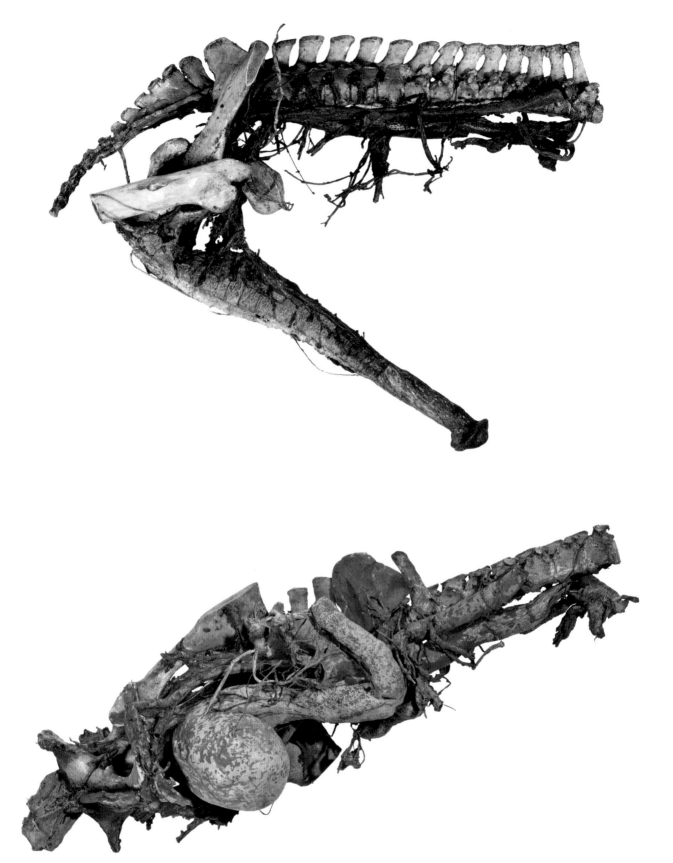

Genital Organs of the Horse

OPPOSITE, TOP

Horse 1203 the organs of generation of a stallion with:

1. the aorta, the mesenteric arteries
2. the veins, the renal arteries, the spermatic cords
3. the ureters, the bladder
4. the vena cava, the hepatic veins
5. the penis, and a superb network of veins along the superior surface formed by the internal and external pudenda,
6. the urethra and the muscles that cover it,
7. the seminal vesicles, the vasa deferentia, the muscles of the penis, the middle vesicle,
8. the part of the thoracic duct that resides in the lower belly,
9. the whole supported by the pelvis and the cervical vertebrae,
10. the tunica vaginalis

OPPOSITE, BOTTOM

Mare 238 uterus of a mare in its natural state, visible are:

1. the pelvis and the vertebral column
2. the aorta, the mesenteric arteries
3. the vena cava, which is very large, the ilia,
4. part of the hepatic veins,
5. the splenics
6. the kidneys and the ureters
7. the bladder and the rectum.
   This piece is injected.

These two specimens illustrate the reproductive system, a group of organs of profound importance to breeding and animal economy. These genital organs of horses have been injected and carefully dissected. The bones have been removed for clarity of presentation, and the long vessels have been reinforced with metal rods to prevent bending.

RIGHT

The Foetal Membranes of the Mare

This orange-colored mass traversed by pigmented vessels is the placenta, or, more precisely, the chorion, of a mare. The arteries and veins have been injected with colored wax and the opening was subsequently closed. The specimen was dried in a distended state to prevent its walls from sticking together, then varnished.

LEFT, TOP

The Lungs

The lungs are probably from a colt or a pony. They were inflated and then dehydrated while still puffed out. A thick varnish coating conceals the details, making identification of the species indeterminate.

LEFT, BOTTOM

The Pig Hearts

Preservation of the heart was not easy, as this hollow organ collapses on itself during desiccation, becoming unrecognizable. Before processing, Fragonard filled it with millet until it was distended and then dried it. The hearts (one shown here) were then coated in colored wax. Grains of the millet are still visible, stuck in the auricle walls.

The *Écorchés* Outside the Museum

Five existing *écorchés* outside the museum are likely the work of Honoré Fragonard.

The Musée Delmas-Orfila-Rouvière has two of Fragonard's specimens, the intestines of a man (actually a child) and a monkey, a species of baboon posed in a squatting position, gripping a tree branch in its hands.

Another *écorché* is preserved in the conference room of the Institut médico-légal de Paris (Paris Institute of Forensic Medicine), on the quai de la Rapée. It is the *écorché* of a woman found in the attic of a hospital in Laval run by nuns and was brought to Paris in 2000. The nuns had attributed the preservation of these remains to the character of saint-liness, which, it is believed, can ward off putrefaction. Christian tradition cites many such miraculous preservations, natural and supernatural. Forgeries and genuine mummies alike are legion in the chancels and crypts of churches and constitute the material support of devout faith. Precise analysis of the components in the injection of the vascular system of this specimen would be required to positively identify its maker, but certainly the injections, the manner in which the cranial vault has been opened, and the dissection of the muscles bear extremely close resemblance to busts at the museum. In outward appear-ance, however, it is quite different from the specimens tidily conserved in the museum's glass cases—this *écorché* remains in its packing case, surrounded by cord and placed on a bed of plain rye straw. It seems to have survived the centuries, slowly accumulating dust that has rendered it dark as pitch. The fine white teeth contrast sharply with the coal-black body.

A fourth *écorché* attributable to Fragonard is one of a child housed in a glass case. It was put up for sale many years ago by an antiques dealer in Rouen

and sold at auction in 2004. Acquired by a private buyer, it now takes pride of place in the boardroom of a business in Val-de-Marne.

Lastly, a fifth *écorché*, by far the most beautiful, is in the collections of the University of Montpellier 2. The origin of this specimen, a little girl presenting a complete myology and angiology, is unknown, but if it is actu-ally Fragonard's handiwork it may have come from Paris when Fragonard was head of preparation of anatomical works at the Paris School of Health. The other French schools were in Strasbourg and Montpellier, and it is likely that this specimen was exchanged between these institutions. Its main object is the dissection of the vascular system, making it an angiology, but the muscles are demonstrated as well. The whole specimen serves to display the vascular-ization as far as the terminal divisions

within the organs. The thorax, wide open, reveals the heart and the great vessels connected to its base. What is remarkable about this *écorché* is that the injections of the arteries have been carried out following the technique recommended by Jean-Joseph Sue; the cannulae were inserted at the base of the aorta and the pulmonary trunk to provide the means of injection of a warmed, red-tinted mixture. Lumps of wax are still visible today where the warm injection leaked on removal of the cannulae. The pulmonary arteries, injected in red, have been painted blue per the standard color code in use in the eighteenth century. The pulmonary veins have been injected in a retrograde manner by forcing open the aortic valves and the left atrioventricular valve; they are painted red.

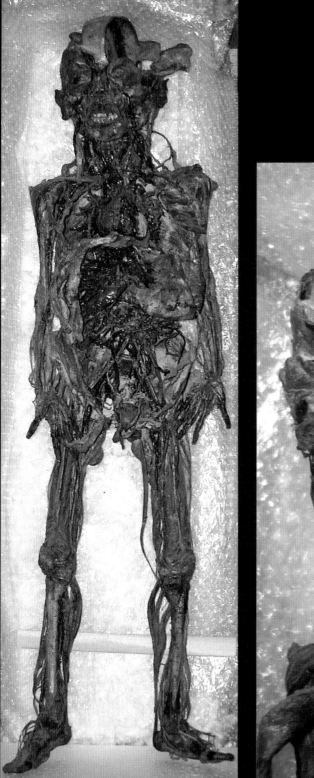

Yesterday and Today: From *Écorchés* to Plastinates

LAURE CADOT,
IN COLLABORATION WITH
CHRISTOPHE DEGUEURCE

Fragonard's *écorchés*, silent witnesses to the history of a discipline built upon the transgressions of age-old taboos, regard us from the recesses of their display cases. Depersonalized by the workings of the scalpel, these fascinating beings today take their place within a well-recognized scientific heritage; they are traces of the milestones in the conquest—as it may aptly be called—and the dissemination of anatomical knowledge since its rediscovery during the Middle Ages up to current-day plastinates. Treatises and manuals, casts and models, dry preparations and fluid-preserved specimens allow us today to understand the intellectual and social construction of the discipline of anatomy, inseparable from the practice of dissection, that arose at the end of the thirteenth century.

Against the backdrop of more than seven centuries of progress in the understanding of the human and animal body, the *écorchés* of Alfort are today seen as emblems of the point when the burgeoning empirical scientific mind at the end of the Renaissance began to merge with a shift inspired by the Enlightenment toward a systematic organization of acquired knowledge. "Between art and science," as many commentators are inclined to stress, the *écorchés* are nonetheless, and perhaps most importantly, the first momentous contributions to the great teaching collections that began to appear shortly before the Revolution and continued to grow straight through the whole of the nineteenth century. Anatomical specimens have always been indispensable in the dissemination of anatomical and physiological knowledge, as well as in education about various scourges to health and hygiene. Since their earliest collection and public display in museums and fairgrounds, they have captivated all strata of society seeking science and the sensational alike. Exhibition display of the time often aimed to offer a diversion as well as to present

Man with a Mandible under protective sheets during the museum's 2006 renovation

an aesthetic experience in the quest to spread knowledge.[1] In the nine-teenth century, in the context of industrialization and social transforma-tion, the hygienists and moralizers, in their concern for the education of the masses, sought to highlight the ravages of alcoholism, venereal disease, and other afflictions (plague, leprosy, cholera, and cancer) and in doing so also helped to popularize science. Despite a popular interest undiminished for more than a hundred years, the turn of the twentieth century nevertheless saw a progressive decline in these anatomical collec-tions, as much on the technical level as on that of the moral, and with a disapproving eye some came to be viewed as inappropriate excesses in displays of morbid pathologies and monstrosities. The advent of increas-ingly accurate and realistic medical imaging[2] and the revulsion with which flesh was regarded in the wake of the atrocities of the First World War gave the coup de grâce to what had become an almost compulsive building of such collections. They began, little by little, to fall into the shadows, leaving the anatomists to pursue their investigations away from the gaze of the uninitiated.

It is important to realize that, over time, these old anatomical speci-mens have largely lost their initial purpose as teaching aids to become instead objects of our heritage. That they remain conserved and available to the general public in various institutions leads one to wonder about the change in how we regard them, and, more generally, about the relation-ship of contemporary society with what can only be called museum pieces. From this point of view, the story of the Alfort specimens, known to be the oldest dried preparations preserved in France, reflects the history of all collections of its type and the hazards they faced. Proudly displayed when new, later they were often destroyed or abandoned for lack of interest or funds, and it is only thanks to the involvement of enthusiasts that a lucky few survive today. The medical students and enlightened aficionados who used to visit the Fragonard Museum have been succeeded by a much more mixed public: families, scholars, historians of science, and artists in large numbers have become enthralled by the museum's atmosphere and the mind-set of an era long gone that is encountered in the aisles of the museum.

In recent years there has been a renewed interest in the history of anatomy,[3] and the development of worldwide shows such as Professor Gunther von Hagens's *Body Worlds* exhibition, and his many imitators, once again puts the unhallowed figure of the *écorché* at the fore of today's news. Is this merely a current fad resulting from the whiff of scandal or is it symptomatic of the corpse's resistance to the dematerialization of our era? The question deserves to be asked.

Today the debate surrounding the conservation and display of human remains has spread far beyond the confines of the collections for study in museums or universities and engages an eager and demanding new public audience. The long queues at the entrance of the anatomical exhibitions that crisscross the globe from Singapore to Los Angeles, making stops in Brussels, London, and Berlin, are testimony to the passion of the public for science—and for a spectacle that is a little macabre.[4] This phenomenon is of course not new and recalls the attraction of the theatra anatomica, the "cabinets of curiosities," and fairground stalls[5] as previously mentioned. For close on four hundred years a mixed public of students, nobles, artists, and ordinary men, women, and children have streamed in to witness scientific demonstrations or to view dissected specimens in an atmosphere that was part social event and part spectacle. With whole bodies positioned in sometimes disquieting poses, lungs blackened by the effects of tobacco, like a magnificent recounting of the past, today's anatomical exhibitions fill the "edutainment"[6] slot, but with a high ticket price. While the visitors of today are numerous as in the past, the evolution of our relationship with the human body, and with illness and death, has noticeably modified our feelings about these pieces.

Reading the comments in guest books from the exhibitions unequivocally shows the disparity of reactions, from total enthusiasm to displeasure and even condemnation. Many visitors appreciate the rarity of such a close encounter with the body and being able to see what has been the reserved domain of medical practitioners. Clarity in the displays is also often mentioned, and the different exhibition rooms likened to chapters in a three-dimensional anatomy book. Still, a certain skepticism about the posing of the specimens on display is sometimes expressed. While the everyday attitudes, such as sporting stances, or artistic references are relatively well received by the public, the reaction was quite the opposite when it came to the presentation of the pregnant woman languidly reclining, her foetus displayed inside her opened uterus, or a couple in full coitus, as in one of the versions of *Body Worlds*. The distinction between a simple pose and a big production, a mise-en-scène that comes across as a great spectacle, depends very much on how the motivation behind it is perceived and how easily shocked the spectators are. For an uninitiated public, the "aesthetic" aspect as distinct from the scientific context is inevitably noticed first. The legibility (which it truly is) of the specimen depends very much on the positioning of the body, which is crucial for putting on visible display the muscles, nerves, and the other vascular systems that are normally concealed. Without proper positioning they would be invisible to the uninformed eye; through positioning the viewer can understand

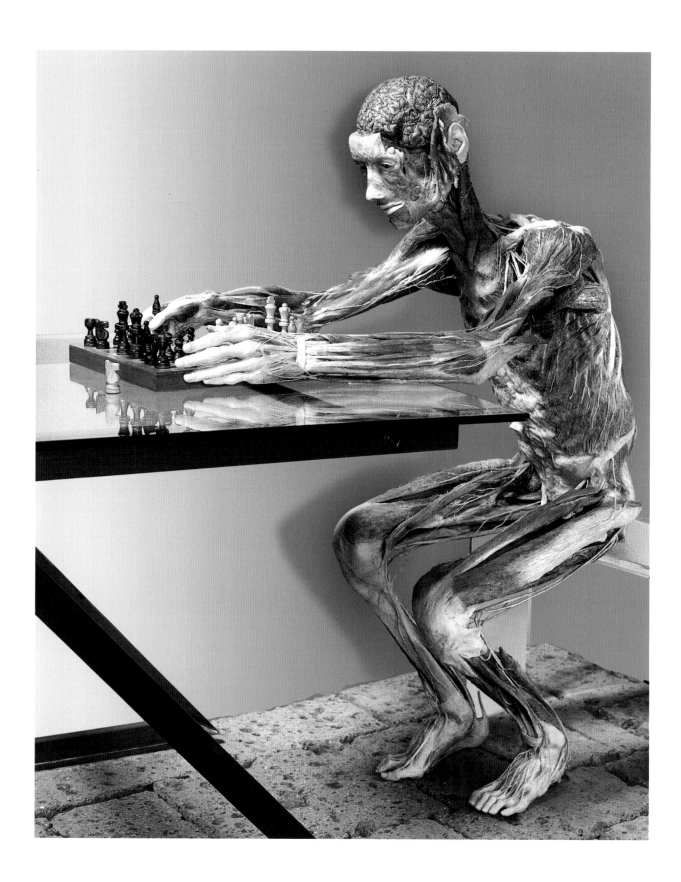

their role within the organism. To give one example, the chess player prepared by von Hagens and copied by other concurrent organizations perfectly presents the involvement of the nervous system in the activity of the subject facing a chess board. His attitude of concentration is a direct reflection of the journey of his thoughts from the brain to the hand, which would move the next piece. Here too the anatomist's guiding principle, which marks his success, is apparent as well: the attitude chosen for the preparation is to serve conveyance of the idea; the aesthetic reinforces the content. Should that relationship become inverted, form becomes more important than the content and the connection between the specimen on exhibit and the manner chosen to display it can rapidly lose sense. Such an outcome naturally devalues any museum specimen and an overstatement of the spectacle can be the result. Lacking an educational purpose, the exhibit soon can become disturbing. These pieces often become controversial and receive a lot of media attention, to the detriment of the discipline of anatomy itself and the understanding of its true goals.

As such, the contribution of plastination, the technique developed at the end of the 1970s by Professor von Hagens, which consists of impregnating the soft tissues with silicone,[7] has undeniably multiplied the possibilities by preserving the volume of the different organs and muscular tissues, a goal never attained with such high quality in the past. Specimens prepared by this procedure can be manipulated, are resistant to decay, odorless and nontoxic, and offer an infinite number of possibilities in their spatial arrangements, guaranteed by the rigidity that the silicone confers. Two centuries of history and technological progress separate the *écorchés* of Fragonard from today's plastinates; despite that distance of time, parallels between the specimens are nevertheless undeniable. Indeed, some preparations, such as the "horseman" produced by von Hagens in the early 2000s and presented like an echo of the one at Alfort, highlight the connection. A distorted view of the work of Fragonard would be to see chiefly the phantasmagorical aspect conveyed by the *Man with a Mandible* or the dancing foetuses. We must not overlook the fact that the primary purpose of these specimens was the demonstration of anatomy during Fragonard's period of teaching at the École royale vétérinaire (Royal Veterinary College). Fragonard was, as we know, neither the first nor the only one to introduce the mise-en-scène into his work. Long before him and in other guises, such as illustrations in anatomical treatises, poses inspired by classical sculpture or religious iconography were reproduced, attesting to concerns about the representation of the body. At a time when there was moral and religious disapproval of, but no actual ban on, the practice of dissection, the anatomist found himself faced—more or less

The Chess Player © Gunther von Hagens's BODY WORLDS, Institute for Plastination, Heidelberg, Germany, www.bodyworlds.com

consciously—with the double necessity of giving back to the corpse some part of its humanity, through memento mori and other scenes of vanity, as well as of exorcising his own uneasiness vis-à-vis that act of transgression, then perceived as a mortal sin. At the same time, under cover of the scientific legitimacy of anatomical representation, anatomists and illustrators of that time had authority to imbue the figures of the *écorchés* with a certain eroticism, as they unveiled their privacy in a mixture of violence and provocation, which is not without invocation of the sadistic imagination of Fragonard's contemporaries. It appears that whether inspired by redemption or by pleasure, the artistic aspect of the mise-en-scène of the anatomical representations has equally contributed to the popularization of knowledge and enhanced the appeal of these specimens to enlightened aficionados, reaching beyond the medical sphere. Despite the need to soften the impact of death naturally omnipresent in any exhibitions of cadaver intended to impart knowledge, the way that the bodies are arranged by von Hagens and his imitators does, however, raise the question of limits and of the respect due to the anatomized individuals displayed. Attitudes about this can vary enormously according to country and culture.

How many specimens today occupy the display cases of the faculties of medicine and departments of anatomy and anthropology in museums that have no such vein of sensationalism? Independent of the losses resulting from the entropy intrinsic to organic materials (deterioration, insects…), the absence of an inspired display as much as the loss of perceived relevance for teaching and the development of new techniques have doubtlessly contributed to the disappearance of a quantity of specimens impossible to estimate but, to all evidence, substantial. Thus, of the thousands of preparations realized by Fragonard, only twenty or so of the most spectacular remain with us today. The history of the collections parallels the history of the discipline, and likewise that of our connection to the body in its most palpable form. So as interest in the collections waned anatomy, a science fundamental to the learning of medicine, has gradually been relegated to a secondary discipline.[8] This situation is reflected today in the increasingly limited opportunities for students to dissect, due to drastic restrictions on the conditions of supply of cadavers as well as the difficulties in ensuring the necessary conditions of hygiene and security. Nevertheless, in places where medical imaging and multimedia have been seen as a providential alternative to cadaveric dissection, in the past several years there has been a "return to the flesh," with prepared specimens placed at the disposal of the students within learning centers in some schools and universities, such as at the faculty of veterinary medicine

Tab. V.   Lib. IV.   17

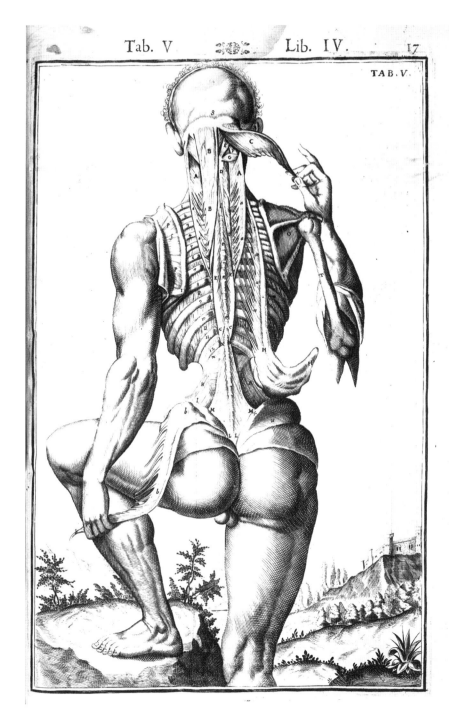

Giulio Casserio, *Tabulae anatomicae LXXIIX*, man with a flayed back, demonstrating the muscles of the back, book 4, plate 5, fig. 17, Venice: E. Deuchinus, 1627. BIUM, Paris

at Utrecht (see p. 81 photograph). It is difficult to pinpoint whether this renewal of anatomy teaching based on the physical study of the body has preceded or followed the rapid growth of the public anatomy exhibitions. The fact remains that, once again, pedagogical anatomy and anatomy-as-spectacle coexist, as in Fragonard's time. It would be simplistic, however, to regard the old specimens in museums or educational institutions solely in terms of their didactic aspect and to see the plastinates that have been displayed in all four corners of the earth as pure spectacle. Like it or not, the curiosity that engendered the *écorchés* remains today and forever at the frontier of scientific interest and of morbid fascination alike, from the point of view of the anatomist and that of the spectator too. What, then, lies behind the controversy that surrounds the revival of these preparations today? Perhaps the ambivalence comes from their being contemporary. Whereas the passage of time distances us from the old specimens, with these new figures we come face to face with men and women of our own time who have lived lives much like ours, before their recent death. Furthermore, the quality of the methods patented by von Hagens, ensuring a conservation nearly perfect, heightens the capacity of the viewer to project onto these bodies his own ills and perhaps even anxieties, indeed to identify with the individual displayed before him. It is difficult not to envisage one's own ultimate fate when faced by the *écorchés* in these exhibitions, especially those that are most imposing and very often openly presented, neither a glass case nor even a guard rail between us and them, while leaflets and information boards about donating one's own body made available at the end of the tour allow everyone to ponder the possibility of one day joining these figures positioned on a bicycle or cut into slices. It is easy to understand that uneasiness might be felt after such a confrontation, especially for people not prepared for such a special encounter with their fellow man.

As for the provenance of the dissected subjects themselves, it is not always as clear as some of the organizers of these events would have us believe. Although death-sentenced prisoners, derelicts taken from hospitals, cemeteries, and the gallows for centuries kept the anatomy theaters supplied, successive laws dating from 1810 in France seeking to outlaw this practice, and the establishment in 1887[9] of a system regulating the donation of bodies for science, decidedly limited the possibilities of supply. Knowing the difficulties that medical schools encounter in obtaining adequate numbers of cadavers for their anatomy classes, one can only wonder about the procurement of bodies for the contemporary *écorchés* displayed outside the domain of a university. To give one example, when the exhibition *Our body: à corps ouvert*, the first of its type to be presented in Paris, opened in February 2009, human rights advocacy groups, suspicious that bodies from executed Chinese criminals obtained in the black market had been used, took the organizers to court; the exhibition was shut down the following April. To date, no judicial inquiry has been carried out, partly because of the difficulties in an investigation with numerous intermediaries, which tends to cloud the question over the origin of the bodies, and regulatory differences between countries. Doubt as to the provenance and informed consent of the candidates for plastination remains. The proliferation of the organizations operating in the lucrative niche of anatomy as a grand spectacle, and the resulting demand for "raw materials," only reinforces these doubts. The commercial value and the colossal success of the first exhibitions have by no means escaped the "followers" of the German anatomist who have profusely plagiarized the concept that has made von Hagens famous. Thus, the promoters of these spectacles today launch themselves into the market alongside their more traditional productions of concerts or more classical exhibitions. Activities that liken the body to a commodity available for purchase, and the specialized organizations that orchestrate their display, raise the question of the status of the cadaver in society and in the law. There now exists a market for old anatomical specimens, which regularly come up for sale, and which change hands for a high price just like collectibles. Anyone can acquire these specimens, now having become objects by the force of circumstances, in essence belonging to no one if not to the family of the deceased, entrusted with his postmortem fate. Obviously such ownership is impossible to identify in the case of anonymous subjects. In the face of this phenomenon, social conscience has been aroused for some time, and ethical reflections on the status of the dead body in the context of heritage are debated. That debate, however, meets a disparity of situations and a lack of clarity in the legal provisions already in place. How can one reconcile, for instance, the laws of bioethics current in France, which stipulate that "the human body, its parts, and its products cannot be made the object of a right of ownership,"[10] and the reality of practices both in the past and today? Should a distinction be made between historical collections, some of which already benefit from the principles of protection established, in France, by those devoted to the so-called musées de France,[11] and those specimens produced by private organizations that do not fall into the public domain? The subject is at the least sensitive and, so far, no single satisfactory response has been made.

Undeniably time, and the difference in motivation, distinguishes medical collections from contemporary plastinates. Our society has become progressively cut off from the proximity of the cadaver since the last public dissections, the end of access to mortuaries in the early twentieth century, and, generally, the increasing isolation of the dying in the hospital. Public opinion reflects this distancing in the new sensibilities that have evolved. Death unavoidably disturbs our daily reality, and not just as a spectacle broadcast on television and on the Internet. Perhaps this state of affairs leads us today to pose the question whether artificial models and synthetic reproductions might be used rather than actual flesh, since this is now technically possible. That alternative, although much used in the past with waxes and plaster models, has never managed

to usurp the practice of dissection and preservation with which it has for so long coexisted—and for a very good reason: modelers and ceroplasticians, while they may be faithful in their reproductions, are not anatomists. Each succeeds to the limits of his or her specialty, and the works created are rarely fully mastered by a single individual. The anatomist remains the sole agent of the ability to display the nature of the body in its physical reality. Similarly, although medical imaging has naturally come to complement the teaching of the discipline, professionals agree there is no substitute for touching and observing flesh while training for medical practice and, in particular, surgery. Disseminating anatomical information to the general public through models would be utterly absurd, an approach suggesting people cannot be confronted with an actual body like their own. This question is part of a more general debate about the use of replicas and facsimiles in place of genuine works of art in museums, which has its advocates, not to protect the public but to protect the works themselves. Such an idea deprives the institution of its very reason for existence, which is precisely to bring the viewer into contact, even if only visually, with the original. What would be the effect on a society cut off from the truth with access only to copies? A refusal to confront the truth would have even more consequence in the case of anatomy. So the unease with bodies in the flesh does not seem to be due to their exposure so much as to a repression that has taken root with respect to the body in its rawest form. At a time when signs of an erosion of standards and values seem to be escalating, a sound return to a physical consciousness of our being through an educational approach permitting a much-needed confrontation with ourselves seems the best route to reminding ourselves of the complexity and fragility of our constitution.

Throughout this problematical story, it is the place of man and the value of his body that are at the forefront. In this debate, the philosophical, religious, scientific, and historic merge as we seek to establish a limit between the diffusion of a knowledge that is of interest to all and the excesses to which the lure of profits or the lack of scruples can lead. Let us hope that everyone succeeds in our shared pursuit so that the controversies of today do not distort the perception of a vital discipline still not well understood by the general public.

Rearing Horse with Rider © Gunther von Hagens's BODY WORLDS, Institute for Plastination, Heidelberg, Germany, www.bodyworlds.com

OVERLEAF
Claude-Florentin Sollier, *The Anatomy Lesson at the House of Lafosse*, c. 1770. ENVA

Appendices

Philippe-Étienne Lafosse, *Cours d'hippiatrique ou Traité complet de la médecine des chevaux* (Veterinary Studies, or A Complete Treatise on Equine Medicine). Paris: Edme, 1772

Preparation Techniques for Dry Anatomical Specimens

Preparation of the *écorchés* was slow, meticulous work requiring extensive experience. Even today, it takes more than three years for a preparator of anatomy to master dissection and the techniques of preservation of the body and organs. Explaining to the layperson the details of these techniques generally seen only in the anatomy laboratory is no simple matter, but it is necessary to understanding the skilled work involved in the creation of *écorchés* like those of Honoré Fragonard. It is equally a necessity for anyone undertaking their conservation to understand the modes of fabrication of these rare specimens. This essay details these little-known methods in two parts, first describing the general methods published by the authors of the eighteenth century, and then the details of the technique used by Fragonard.

Classical Techniques in the Creation of Écorchés

The anatomist needed the body to be very fresh, so it was retrieved as soon as possible after death. The choice of cadaver was key: best were the bodies of young persons, with a flexible vascular system, and thin, to make dissection of the fatty regions easier. In the workshop, the body was washed in warm water and closely shaved before the superficial blood vessels were cut open to expel thickened blood. The anatomist could inject warm water or even air into the vessels to make this easier, but doing so brought risks to the later injection of resin mixtures, and injecting rinses was practical only in very fresh bodies. The body was then warmed in regularly renewed hot water, and after several hours it was ready for the injection of the mixtures to fill the vascular system. The anatomist made a longitudinal cut into the skin and subcutaneous fat along the length of the sternum and then extended the line of the cut by two incisions along the lower costal cartilages. The skin and the fat were dissected away to reveal the pectoralis major muscle, which was detached from the ribs and their cartilages. The ribs thus uncovered were sawn through to open the thorax. Immediately visible was the pericardium, into which the anatomist made a cross-shaped cut to reveal the heart and, at its base, the pulmonary artery and aorta. These two arteries were separated one from the other by means of a curved needle or the tip of a pair of dissecting forceps. Three or four strands of waxed thread were looped around the aorta a short distance from its origin, and the arterial trunk was incised above the ligature. The blood discharged from the opening

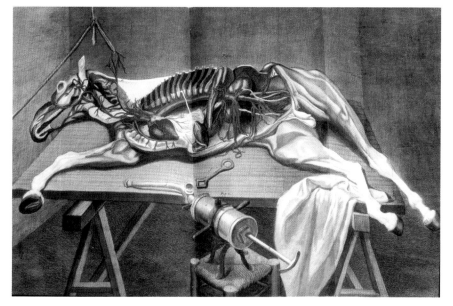

Angiology showing the aorta and its branches. Philippe-Etienne Lafosse, *Cours d'hippatrique* (Veterinary Studies), plate 25. Paris: Edme, 1772

was cleared to allow the insertion of a tube that was firmly tied in place: it was important that the pressure of the injection not push the tube out of the aorta. A similar injection could also be made in the thigh or the arm. The injected substance was forced into all the arteries with a single stroke of the syringe's piston, with the exception of the pulmonary and coronary arteries, which had to be treated separately. Success of the operation depended on the composition of the injected material.

The Injection

Jean-Joseph Sue, one of the great anatomists of the eighteenth century, revealed his methods.

He used two successive injections, one fine and one coarse. The fine injection was made with the aid of essence of turpentine, into which he mixed a pigment, red for the arteries and blue

for the veins: "Two or three ounces of the very finest quality[1] vermilion are placed in a clean pan or glazed earthenware pot; add to it several spoonfuls of essence of turpentine, taking care to stir well with a horsehair brush to dissolve the vermilion. Up to about one pound, more or less, can so be made depending on the quantity of the fine injection that one wishes to make, or rather, the size of the subject to be injected. The solution is then filtered through a fine cloth to eliminate any large particles, allowed to rest for a few minutes, then poured into another clean vessel. This filtering is repeated three or four times until no more large particles remain on the cloth."[2]

Sue's coarse injection was prepared in the following manner: "One must have available an earthenware pot, glazed and very clean, to be placed on a low fire. Place into the pot one pound [490 g][3] of mutton or beef fat and six

ounces [185 g][4] of yellow or white wax. Melt them together, stirring occasionally with a spatula or horsehair brush. When completely melted, add three or four ounces [90–120 g] of lard, or three ounces [90 g] of olive oil and four ounces [120 g] of turpentine of Venice."[5]

The pigments could be animal, vegetable, or mineral in origin: "When the materials are well mixed, add three ounces of vermilion for red or for blue, add three ounces of verdigris or Prussian blue, crushed in oil in a bladder, gradually thinned with the liquid in another terrine of which the bottom should not be heated. When the whole is well mixed, pass the liquid through a slightly coarse, white cloth to remove any large particles; provided that it is suitably warm, it is ready for injection."[6]

The coarse mixture, injected after the fine injection, forced the latter into all the smallest capillaries while filling the large vessels with its bulk. The essence of turpentine, being very volatile, evaporated from the red dye remaining in the small arteries. Let us imagine the anatomist at work, at the decisive moment of the injection. The body, warmed "to the heart," is placed on the table. The barrel of a syringe or piston pump is filled with the first mixture, the fine injection. Trapped air is expelled from the instrument; otherwise the injection would fail, as any air bubbles in the vessels would block passage of the liquid, producing colorless areas. The syringe is then connected to a tap that is fixed to the tube introduced into the aorta. Slowly the "liquid" is steadily pushed by depressing the pump until all has been injected. The tap is closed to prevent leakage and the syringe removed. The coarse injection, already prepared, lies in wait warm enough to flow through the vessels without setting. The syringe is refilled, cleared of air, and fitted to the tap. Swiftly but smoothly, the anatomist forces the preparation into the vessels until a slight resistance indicates that

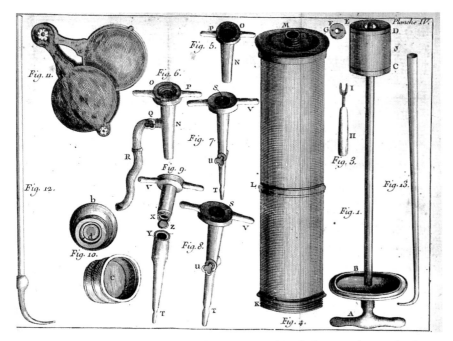

Injection apparatuses. Pierre Tarin, *Anthropotomie, ou l'Art de disséquer les muscles, les ligamens, les nerfs et les vaisseaux sanguins du corps humain…* (Anthropotomy, or The Art of Dissecting the Muscles, the Ligaments, the Nerves and the Blood Vessels of the Human Body…). Paris: Briasson, 1750. ENVA library

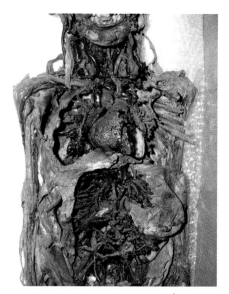

Attributed to Honoré Fragonard. Écorché *of a Young Girl* (detail). Visible at the base of the heart is the injection that leaked out when the cannulae were removed from the aorta and pulmonary trunk. Université Montpellier 2

the arteries are full, and the cannula is closed with a half turn of the tap to prevent leakage of the mixture. For subjects such as large animals, several syringes had to stand in readiness, or several quick refills of the syringe would be required. It took at least half an hour for the mixture to cool and solidify before the body could be wiped clean and the dissection begun.

Certain arteries such as the coronary and pulmonary required a special injection. For the coronaries, a small tube could be placed at the root of each artery, fixed firmly in place with a ligature before the injection, or the injection could be carried out directly via the left ventricle, with the tube placed right at the base of the aorta.[7] The injection of the pulmonary artery was no more difficult; all that was necessary was to repeat the same operation, placing a tube fitted with a tap in the vessel or in the right ventricle. Caution had to be taken not to use too much force or the pulmonary

vessels were at risk of rupturing and the injectant would spread into the pleural cavities.

Injection of the heart itself was easy. Two tubes were fitted, one at the root of the pulmonary trunk and the other at the root of the aorta, directed downward, through the valves to the ventricles. The compound was pushed in gently, a little at a time, to allow the fibers of the heart to stretch, or else the wall of the ventricle—particularly the right—could be torn and the specimen ruined.

The Foetus: A Special Case

The museum has several injected foetuses, and these pieces are by necessity prepared by a different technique, as the circulation of the blood does not occur in a foetus as it does in an adult because of two communications, one between the pulmonary artery and the aorta and the other between the two atria of the heart. If only the arteries were to be injected, the ductus

Honoré Fragonard, *Group of Three Foetuses* (detail), 1766–1771. The arteries and veins of the umbilical cord have been injected and the vascular network of the liver is displayed.

arteriosus was ligated to prevent the liquid from passing from the aorta to the pulmonary artery, into the right ventricle, into the right atrium, and into the veins. If one wished to inject both the arteries and veins of a foetus without opening the chest, the liquid was pumped through the umbilical vein: in this way the inferior vena cava, the portal vein, the right side of the heart, the pulmonary artery, and the aorta via their communication, and then all the arterial system was filled.[8]

Injection of the Placenta[9]

Although placental vessels are clearly visible, injection via the umbilical arteries is rarely successful. The best method consists of choosing one of the larger branches at the surface of the placenta and proceeding with the injection first in one direction, then in the other. The veins can be injected through the umbilical vein.

Injection of the Penis[10]

The corpora cavernosa, the urethra, and the glans are very elastic and capable of much expansion. First, the anatomist injected the arteries. To do this, if they had not already been filled via the aorta, he placed a tube near the origin of the internal iliac arteries and pumped the mixture through it. He then moved on to injecting the corpora cavernosa: having located one, close to the ischium, he made an opening large enough to introduce a tube and, using a stylet, destroyed the internal septa. The injected substance could then spread through the penis and into its veins. Next he located the bulb of the penis and again ruptured the septa with a stylet before injecting. If the injection failed to reach the glans, an injection was made directly into its ventral surface, close to the frenulum.

Injection of the Veins[11]

Injection of the venae cavae was made easier by the wide caliber of these vessels. Veins of small diameter were much more difficult and laborious than the arteries. The presence of valves directing the flow of blood from the extremities toward the heart necessitated multiple injections at the periphery of the body. In the hand, for example, each digital vein was individually catheterized and injected.

Displaying the Lymphatic System[12]

One of the greatest challenges to display was the lymphatic system. Little known by the general public, it represents the third component of the vascular system, after the arterial and venous networks, and transports lymph, which makes up 10 percent of the circulating volume. It is very difficult to observe because of the delicacy of the vessels, and the lymph that they contain is transparent and colorless. To visualize these vessels, the anatomist had to use a trick, taking advantage of the milky appearance that the lymphatic vessels from the intestine assume following a fatty meal. The animal was given a meal of milk or fresh cream before it was sacrificed. The abdomen was opened to spread out the intestines and their mesentery, in which the lymphatic vessels appeared white— the injection, although difficult, became achievable. The cysterna chyli and the thoracic duct could be injected directly,[13] even though the thoracic duct can be sometimes very difficult to find. Once the anatomist had located the duct he introduced a narrow metal tube into it and blew air into it so as not to lose sight of it. If the dilation showed that the tube was correctly positioned, he injected a compound not too thick and white in color and allowed it to cool before turning the tube to proceed with a retrograde injection.

A Method to Reveal Injected Vessels: Corrosion

Injection also permitted corrosion of the body to reveal only the vascular system. The principle is simple: a resistant material is injected into the vascular system, and the body is exposed to a corrosive "digestive" so that nothing remains but the cast of the arterio-venous system. This type of specimen, particularly useful for teaching, demonstrated the finest vessels, preserving them as long as the specimen escaped destruction by insects. It required skill and care from the anatomist. The substance used in the procedure had to be fluid enough to penetrate the smallest vessels but neither melt in summer nor become too brittle in winter. Also, the coloring pigments must not be affected by the corrosive *menstrue*,[14] that is, the solvent used to dissolve the flesh. After injection, the body was submerged in a bath of this corrosive liquid, and as particles of tissue were loosened by the acid they were delicately removed by the anatomist, with either a scalpel or a trickle of water. The classic injection used in corrosion casts was a mixture of tallow, wax, turpentine, oil, and other ingredients, which combined were not resistant to the corrosive solutions, making it necessary to use high-quality white or yellow wax mixed with purified resin and "spirit of turpentine."[15] The resin was melted over a low fire before being strained through a cloth to remove foreign bodies that were often found in it. It was then mixed with the wax, which had been gradually melted and also passed through a cloth; the coloring pigments were then added, and finally the spirit of turpentine, which liquefied the mixture. Although this mixture could penetrate the finest vessels, it had the disadvantage of becoming brittle, so that after the corrosion the finest vessels were easily broken. The celebrated John Hunter[16] used a composition with a resin base, white wax, and purified turpentine of Venice.[17]

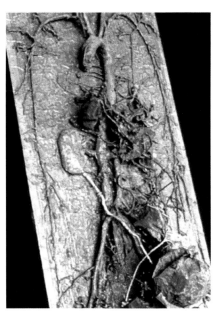

Honoré Fragonard, *Corrosion Cast of the Vascular System of a Woman*, 1766–1771

The corrosive solution was usually spirit of *sel fumant*,[18] used pure, or *l'eau forte*.[19] The organ was submerged in the liquid for several weeks, then cleared of tissue degraded by the corrosive. A difficult maneuver!

Myologies, Neurologies, Angiologies

Once the injections were completed, the cadaver—human or animal—was dissected. Specimens demonstrating the muscles separated were listed by Fragonard as myologies. Two other preparation types were even more delicate. A neurology aimed to show the nervous system: brain, spinal cord, and nerves. The difficulty of this work can be imagined. Nervous tissue is very fragile and decomposes rapidly. The anatomist had to work with haste, increasing the risk of cutting one of the fragile components, and with nothing but alcohol baths to retard decomposition. The peripheral nerves were carefully exposed with a scalpel or scissors, and the brain and

Honoré Fragonard, *Human Bust* (detail), 1766–1771

the spinal cord were released from their bony casing, which had been previously cut or broken away. A dissection displaying the heart, the arteries, and the veins was called an angiology. Scissors were preferable to the dangerously razor-sharp scalpel for creating the angiology. The vessels of the mesentery required particularly patient removal from the fragile intestines.

Opening the Cranium to Reveal the Dura Mater[20]

The brain is composed of very fragile tissue with a high fat content, making it very difficult to preserve in the absence of formalin. Fragonard had of course no such substance at his disposal. Like the ancient Egyptians, anatomists were constrained in how the brain could be removed to reveal the intracranial structures. First the scalp was removed, or peeled back with its underlying connective tissues, then after making a circular-saw cut in the cranium the skullcap was removed. The dura mater exposed, it could be prepared by making small openings to extract the brain, thus creating a cavity that was stuffed with horsehair. Fragonard's *écorchés* have revealed two additional techniques, visible on the two busts. In the first trepanning was used to extract the brain so that the head could be preserved; the second shows

the falx cerebri and the venous sinuses that it contains. The cranium was sawn so as to leave an osseous "bucket handle" supporting these folds of the dura mater.

"The Art of Conserving the Prepared Parts, Fresh or Dried"[21]

When the moment arrived to conserve the finished specimen, it could be placed in a preservative liquid, generally alcohol, into which could be mixed a great variety of additives. To do this, the specimen had to have been previously emptied of blood by frequent washes in cold water. This procedure maintained the organ's volume and flexibility, maintained its original appearance, and allowed further work later on.[22] Each anatomist had his secret methods. Frederik Ruysch, for example, who excelled at this technique, used a "balsamic" liquor that he maintained was made from *eau-de-vie* distilled from cereal grains, into which he crushed white pepper; some anatomists have reproduced the procedure—without success. And Alexander Monro added nitric or hypochloric acid.[23]

Preservation of specimens by desiccation required yet more steps. The fresh specimens were first submerged for eight to fifteen days in alcohol, before being soaked in a solution of vinegar spirit,[24] to which was added "corrosive sublimate," a product that protected against insects.[25] They were then arranged in the desired position, the aim being to give to the specimens, both whole bodies and anatomical fragments, a lifelike attitude. To achieve this, the subject was placed in a frame onto which threads, pads of horsehair, pins, pieces of card, or small sticks were attached to hold the muscles in position. When secured, the specimen was exposed to air at a moderate temperature. As it gradually dried, the preparator had to reposition the muscles and other parts to avoid any shrinking and deformation that would detract from

its usefulness for teaching. The supports were progressively removed and, when totally dry, the *écorché* was coated in a white varnish,[26] generally alcohol based. This stage, long and difficult, aimed to highlight the organs by coloring them. The varnish, originally white, could be pigmented with carmine to restore to the muscles a flesh color. The arteries were painted red with vermilion and the veins blue, with either ash blue[27] or Prussian blue[28] pulverized in oil and mixed into the varnish. The nerves were painted with white lead, which was dissolved in the varnish.[29]

Finally, to ward off attack from insects—larvae or adults—the *écorchés* had to receive regularly coats of alcohol or oil of turpentine,[30] particularly where infestations had already started. This was especially important between January and August, when the larvae of insects that would decimate the collections were developing. To guard against insects and mice, Alexander Monro recommended "corrosive sublimate" mixed with spirit of wine,[31] and in the nineteenth century preparators began to use excessive amounts of arsenic, a powerful poison that today contaminates most of the anatomical specimens and taxidermic animals prepared in that era.

The Secret of Fragonard's Injections

As we have seen, Fragonard did not achieve in 1792 the anatomy museum he

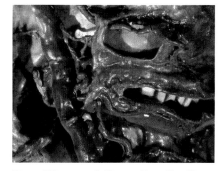

Honoré Fragonard, *Human Bust* (detail), 1766–1771

had pinned his hopes on, and he never divulged his own techniques. It would have remained a complete mystery for even longer if unusual circumstances had not led to a study of the injections and the varnish that he used. Although the *écorchés* had remained in a relatively good state of preservation up to 2003, they were damaged in the heat wave that swept France that summer. The vascular injections started to melt and drops of "wax" dotted the floor of the display cases. Was this something new? Not according to those who had known the museum for several decades. One former employee recounted that during hot summers the *écorchés* "sweated wax." He pointed to a mark left by a glass dish he had seen in place a long time ago under the great vessels of one of the thoraxes. The dish had gradually become filled with the substance of the injections, which dripped from the

Endoscopic study of the thoracic cavity of *Man with a Mandible* carried out with Ron Beckett and Jerry Corlogne of the University of Quinnipiac.

aorta and the superior vena cava. Thus this phenomenon was not recent; the injections had been slowly melting—probably since their creation. A thorough examination of the *Man with a Mandible*, for example, revealed drops of melted wax at the extremities of the mesenteric vessels. The pulmonary arteries and veins had collapsed in the thorax, and the heart was coated with wax that had progressively oozed from the surface. Similarly, in the *Horseman*, the internal surface of the sternum is covered with a large mass of wax, which has probably dripped from the vessels of the thorax and is slowly seeping between the pectoral muscles before falling to the floor, to give some idea of the extent of damage.

The desire to save these unique objects led to a multidisciplinary effort aimed at discovering Fragonard's method of preparation, and thus toward proposed measures for conservation. The opportunity came with a request for authorization to film by the National Geographic Channel, as part of the television series *The Mummies Roadshow*, a documentary on the work of two scientists little by little discovering the secrets of mummies. An agreement was reached whereby they would conduct an investigation into the *Man with a Mandible*, a specimen of excellent technical quality and spectacular vascular injections, in which tests would be made on its injection material and varnish. The tests were carried out at the Laboratoire de Recherche des Monuments Historique, and results were obtained from the Laboratoire d'Etudes des Techniques et Instruments d'Analyse Moléculaire of the Institut Universitaire de Technologie de Paris XI-Orsay, verifying and completing the analysis. The investigation was conducted in several stages: physical examination, radiography, endoscopic identification of pigments, and analysis of the injected substance and, lastly, of the varnish. It has revealed many of Fragonard's secrets.

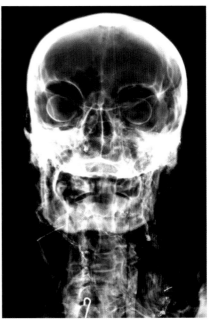

Frontal radiograph of the head of *Man with a Mandible*. The porcelain eyes are clearly visible within the orbits, as well as the vessels injected with radio-opaque products and the needles that held them in place.

All the cadavers have undergone a thoracotomy. In the humans it was carried out according to classic methods; in the animals, the thorax was opened from both sides, revealing the heart and pulmonary vessels. None of these subjects presented with an incised aorta or vena cava that might suggest they had been catheterized. Radiographic studies have shown that the heart was injected in both the left and right ventricles, so it may be that the cannulae were positioned directly in the ventricles. This is immediately visible in the nilgai, in which the heart presents at its apex a purse-string suture made with horsehair. This procedure probably gave the anatomist greater force of injection. The substance injected into the left ventricle penetrated the aorta and spread throughout the arterial network, while that in the right ventricle entered the pulmonary arteries.

Radiographic examination of the cadavers revealed that different compositions were used in the arteries and veins. For example, the radiograph of the pelvis shows a metallic density in the iliac artery, while the neighboring vein is radiolucent. One can just see that the wall of the latter is underscored by a metallic line. Therefore, the arteries were injected with a substance containing a metallic salt, which is not found in the veins. All depends on the pigments. Endoscopic examination of the body cavities showed clearly that the interior of the arteries is stained red while the contents of the veins are simply brown, not pigmented; the injection here is crude, unstained. Analysis of the red pigment used for the arteries showed that it was synthetic vermilion, a mercuric sulfide,[32] which accounts for both the very bright color of these vessels and the presence of metal detected in the radiographs. Moreover, as Sue had recommended, all the vessels were painted: the arteries in red with a mixture including vermilion, the veins in blue by means of a pigment that was found to be azurite. This difference in the pigmentation of the arteries and veins is easily explained: because the veins have valves, their resistance to the injection is greater than in the arteries. The anatomist did not therefore risk making the mixture more viscous by adding pigments. The difference in staining was sufficient to allow easy distinction between the arteries and veins during dissection. The vessels marked with red were painted bright red at the end of the dissection, while the wide, thin-walled veins were identified and painted blue.

More surprising was the discovery of the composition of the mass of the injection. Whereas it had been expected to contain wax, instead mainly mutton tallow mixed with pine resin and an essential oil were found. Only the large animals, the horse and the nilgai, showed traces of beeswax. None of the human cadavers contained it. While this formula may be very close to that of Sue, the absence of this essential component explains the ease of use of this technique. Mutton fat melts at a low temperature: 50 percent of its mass is liquid at 68° F (20° C), in contrast to beeswax with a melting point around 158° F (70° C). It was thus easy for Fragonard to melt the mutton fat and add the pine resin and an essential oil to it[33] to obtain a fluid mixture easily injectable into a warm corpse. In contrast to conditions required with Sue's formula, Fragonard did not need to warm the corpse overmuch to give his injections a greater chance of success than with the conventional method. We can now better understand why Karl Asmund Rudolphi marveled that Fragonard succeeded six times out of ten. During the cooling of the mixture, the resin thickened the tallow, which then formed into a hard mass. Fragonard also avoided the problem noticed by Monro[34] in the case of injections using melted tallow alone or mixed with oil of turpentine, i.e., the resulting fragility of the vessels, which break when handled.

This technique was revolutionary and it can be found in nineteenth-century literature. It was Jean-Nicolas Gannal (1791–1852) who first published in 1838 a similar formula,[35] duplicated by his disciple Pierre Boitard[36] (1789–1859) and printed in the *Grand Dictionnaire du XIXᵉ Siècle* by Pierre Larousse in the article "Injections."[37]

The Dissection

Once the body had been injected, Fragonard proceeded with the dissection according to the classical method. First, the skin was removed. If it was an animal, the cutaneous muscles of the trunk were left in place. Each muscle was then isolated, while preserving the nerves, arteries, and veins. Organs that were difficult to preserve, such as the lungs, the intestines, and the brain, were removed. In humans, two different treatments could be used for the eyes. The first consisted of preserving the natural organs and inflating them; the second, as can be seen in the *Man with a Mandible*, consisted of removing the eyes of the cadaver and replacing them with artificial, porcelain eyes. The appearance was obviously very different: the natural organ in place looked glazed, but the second method resulted in a clear, bright, and rather disturbing look.

Positioning and Drying

In the preservation phase, the mummification following dissection, the body was submerged in alcohol, as is evident from the presence of fatty acid ethyl esters in the injection. Once it was saturated, the body would have been removed from the tank and put into position. This must have occurred in the drying room, a room located above the dissection room that Fragonard mentioned in his 1794 inventory, which also contained the pathology collection. The *Man with a*

View of the sacrum; three holes allow the body to be fixed to a plank and show how the vertebral canal was opened to permit insertion of a metal rod to stiffen the trunk.

Mandible was then posed in the position desired. But how to keep a dead man standing? The answer is still visible. The sacrum is pierced by three screw holes and the vertebral canal open where a metal rod was introduced. The back was rigid; the sacrum was fixed at normal height and the legs were stretched downward to the floor. Fragonard probably used a frame, in which a variety of means were available to hold the head and arms in position. The muscles and the vessels were held in place by needles, some of which are still visible at the exterior of the cadaver, while others more deeply placed are clearly seen in radiographs. The needles were inserted in the body before the vessels were painted and the whole cadaver varnished, as is evident from the successive layers of coating that were applied to the body. As Sue explained, Fragonard constantly had to reposition the anatomical structures as they dried. One can imagine the anatomist at work on the ears and the lips of the man to give them the aggressive attitude still so striking today. The same type of procedure would have been adopted for the *Horseman* and his mount.

Painting the Vessels and Varnishing

In the final phase of preparation, the arteries and veins were painted. One imagines that this took place when the body was already positioned. Indeed, the needles holding the muscles and vessels being in place suggests that the *écorché* had been posed in its final position, with the needles placed to resist the sagging effect of gravity on these structures. While the work is remarkable overall, in spite of great care Fragonard left traces of paint on the anatomical structures near the vessels. Thus the biceps brachii of the left arm has traces of blue paint, and the endoscope reveals traces of red and blue paint on the surface of the interior of the thorax. Once the body was ready, it was varnished with a mixture containing turpentine of Venice, a very pure resin produced from the larch tree. This costly substance was used in the manufacture of varnishes for paintings. It has the major advantage of yellowing without blackening—the colors are still clearly visible on the *écorchés* in the museum—and its plastic properties make it highly resistant. Despite its high price, it was this resin that Fragonard used to safeguard his *écorchés*. And it is probably owing to this varnish that these pieces were protected from attack by insects and survive yet today.

Detail from the arm of *Man with a Mandible* showing traces of blue paint mistakenly splattered on the muscles nearby the veins.

Fragonard's 1792 Report and Notes Addressed to the National Assembly

Arch. Nat., F17 1318, file 4, July 1792. Annotated: sent to the Comité d'Instruction publique (Board of Public Education) July 21, 1792, F.-M. Caihasson, secretary. On the sleeve that contains two reports, in contemporary handwriting: * NOTE. The inventory of the specimens forming the first collection is missing.

The nation of France possesses nearly everything that is needed for all the sciences to flourish, except for anatomy.

She has, in fact, two cabinets one of which, located in Charenton, is too far from Paris, and contains mainly veterinary anatomy. The second, in the Jardin du Roi, displays only objects of pure curiosity, and its anatomy collection is limited to general natural history.

There are, on the other hand, various anatomical specimens in the schools of surgery and medicine, but these collections are too incomplete to be of any real use, and their imperfections show the laudable, but insufficient, efforts of those who have had the patience to prepare them.

Thus, except for a few small "cabinets of curiosities," all Mr. Fragonard's handiwork and privately owned, we have nothing, absolutely nothing in France to shed light on the wonders of that subject, which, despite so much hard work, has not much progressed.

We wish to propose to the National Assembly the establishment of an institution that we do not currently have, an institution without which medicine and surgery will remain at a standstill, the establishment of a national anatomy museum.

Foreign nations, and principally England, which recognize the necessity of the institution that we have the honor to propose to you, have hesitated not even an instant: several millions placed at the disposal of the famous Hunter have given to the English nation one of the most beautiful collections of this type that one could wish to see. But without the operating funds, the specimens will deteriorate and will no longer be useful to the sciences.

M. Fragonard* and Mm. Delzeuzes and Landrieux, his two collaborators, present themselves to carry out preparations for the proposed museum.

Even before the Revolution, which has given liberty to France, it was their intention to combine their fortunes for the establishment of such a museum wherein all the scholars of Europe could access anatomy in all its divisions, man and animal alike, of the highest possible degree of perfection; wherein one could encounter everything needed for study or for reaching even greater discoveries to alleviate the suffering of humanity; wherein the nation's teachers could lead their students to absorb the innermost secrets of the body, especially during the warm season, when dissections and consequently anatomy courses are completely impractical.

So it is that they have worked on a great number of preparations in their own possession, which, aside from the advantage of being unequaled in Europe, owing to their unique injection formula, which others would in vain wish to purchase, are blessed with the property of being unassailable by the worms that can destroy all that one has been able to create. And, indeed, I, Fragonard possess some specimens thirty years old and that are as beautiful as on the first day they were created.

But a change of circumstances since the Revolution has deprived them of the funds allocated to this enterprise.

Do not believe, gentlemen, that it is necessary to follow the example of England and dedicate three million to this goal; our patriotism would not permit us to ask such an enormous sum in today's circumstances. Still, something must be done.

We have calculated everything and we have found (and you will be astonished) that twenty thousand livres is sufficient to found and maintain the richest, most useful museum in Europe. We will in fact supply, and free of charge, about 1,500 specimens, at a value of about twenty thousand livres; upon the most detailed scrutiny, it will be seen that these are specimens impossible to find elsewhere, unless they have been taken from the hand of Mr. Fragonard or of his collaborators.**

Here, gentlemen, is an opportunity for you to see: first, that the Nation will immediately possess, before even providing the funds for the first year, these 1,500 specimens worth at least 20,000 livres; second· that the when museum is complete, when it contains at least ten thousand specimens for demonstration, the Nation will have an ownership valued at more than three million*** and that this institution, considered as a financial operation, therefore is of a very real and very considerable benefit. You may perhaps object that what we offer has no redeemable value; in response we would suggest that nothing would prevent the Nation from holding a sale of the entire collection every five or six years, as soon as new pieces are ready to replace the originals, and that such a sale would even be valuable in consideration of the great usefulness of these specimens should they be dispersed throughout France and the whole world.

Finding a suitable location will present little difficulty: the Nation now owns a quantity of buildings superb and spacious: would it not be a terrible loss to architecture if the Dome of the Assumption, for example, were to be sold, and consequently demolished (for what would such a vast dome be suitable for?); rather, we envisage that majestic temple surmounted by the emblem of anatomy, armed not with a scythe, but with a sacred hat, showing to future generations that Liberty, although still in its infancy, also knew to honor the arts. By its disposition, this location needs very little new—a flight of stairs, two floors, and some display cases. Below would be an amphitheater; on the first floor, the collection; and under the vault, the laboratory.

Ten thousand livres is all that is needed; the two small wings of the dome will suffice for the lodgings of Mr. Fragonard, his two collaborators, and a custodian and museum assistant.

The use of the twenty thousand livres granted by the nation:

Mr. Fragonard, director	2,400
M. Delzeuzes, inspector of the specimens	2,400
M. Landrieux, secretary and keeper of the cabinet	2,400
A custodian	800
A museum assistant	800
	8,800 livres

A budget shall be held for the three undersigned anatomists for expenses for instruments, materials, office sundries, lights, wood, etc., which shall amount to a total of not more than
the sum of 1,200,
bringing the total sum to 20,000
not to be paid by the public Treasury until municipal governance has approved the allocation. It goes without saying that the principals of the anatomy museum must be authorized to choose in the hospital the subjects that will be most useful to them.

   * Mr. Fragonard, creator of the cabinet at Alfort near Charenton and of the most valuable specimens in various collections in Paris.

  ** It will seem less astonishing that for such a modest sum such a vast museum can be created if you keep in mind the simple fact we possess a quantity of very valuable items that would be placed at your disposal.

*** See the attached general inventory.

Addendum

Some days after the memorandum we had signed was sent to Mr. Taillefer, we had occasion to read the Report of the Board of Public Education by M. Condorcet.

With pleasure we saw that they recognized the need for an anatomy museum. It appears to us, as it had apparently also appeared to the board, since it went unmentioned, that the individuals with the necessary skills were not in question in the broad overview presented in the Report, as they must needs be involved with various collections. No doubt the board reserves the right to consider each objective separately, after the National Assembly has decreed the general principles.

Because our wishes are aligned with public interest, and because our wish is to be in agreement with the noble goals of the board, we have formulated some reflections that we now have the honor to lay before the board. It will be seen that although these reflections relate to our business, they should, however, not be received as demanding an exception to the general rule for us alone, for they apply as well to all individuals engaged as chief of collections, even if less directly, in that whereas anatomy requires the most exacting work possible only through the most stringent, precise, and thorough training, collections of all other types demand only a skillful laborer who may be purely passive. Permit me to explain.

We do not believe that a teacher is able simultaneously to be a manual worker; his lessons occupy him, and if he is conscientious, they take up all his time.

For the same reason of a lack of time, an individual who capably prepares demonstration specimens cannot occupy the post of professor.

To view this individual as a laborer is untenable, for apart from the fact that such a position is not secure enough for a man to want to sacrifice a part of his life and probably his health, it would also be unacceptable for an educated man to be subordinate to another man *less educated than him* within the same art,[1] working less time and in less danger; is it not to be expected that the working anatomist would be disturbed by the teaching anatomist, to whom he would be subordinate? Should the teacher replace him with a student? No, gentlemen, the door that would open to this abuse must be absolutely

closed, for the public would not tolerate the injustice that would follow; good anatomists are not as common as you might think.

We therefore believe it essential that the individuals required for the formation of a museum should be named as curators of their own work, and for the Nation to avoid redundancy, one worker can carry out several combined roles, with a constant eye toward maintaining value.

Such a post, secure, honorable, and protected from all foreseeable disadvantages, would benefit the anatomists.

Why, it may be asked, are there three of us? In the interest of the Nation we must not disclose our methods. It may be necessary for one or two of us to be absent to create collections in other *départements*, especially specimens that cannot be transported without risk of damage, and it is imperative that our methods not be divulged, for if all the world knows, then the specimens will consequently have no value, and the Nation's investment henceforth will be destroyed. It is therefore impossible that one alone could do everything.

A similar objection may be raised contrariwise, and we must admit it is not without basis, that on the other hand it is impossible for but three men to create enough specimens for all the many *lycées* (high schools) of the Empire, and what if our methods were to perish with us. To that we readily respond with an offer to train for free the students who help us, and thus the most able from among them will be initiated into the practice and will succeed us, and that arrangement shall remain as long as the Nation judges it necessary to keep the methods private to herself.

Furthermore

The difficulties that the National Assembly suffers at this critical time suggest to us that it will not long be part of the Organization for Public Education. Must we thus remain in suspense? Shall we not begin work on these works that take so long to perfect, until after the final decision? We would not then be able to have the other specimens ready in time.

We therefore ask that the board, after having verified our statements as to the perfection in quality of our preparations and of our injections, presuming they judge them to merit the claims that we make for them and are persuaded of their usefulness,

> willingly propose to the National Assembly that we receive authorization to begin work, on behalf of the Nation, on the subjects that the hospitals hand over to us;
>
> give us a temporary location for working and storing the specimens; in which case we shall donate to the nation the existing pieces at an agreed price;
>
> ensure that our expenses be scrutinized by the commissioners or by the whole committee, and that the method of payment be established;
>
> ensure that our provisional fees be paid by the National Assembly, except as modified or added to as and when the assembly rules on the salaries of public servants in service of teaching, or fixes that of the conservators of the high schools or other educational institutions.

1. That the professor is less knowledgeable than the anatomist is a fact in need of no proof. The one cannot do the work of the other just because he has been the other, and while the professor configures his own theoretical physiology based on that which he may or may not have seen, while he reads very eloquently from a text, it is undeniable that his faculties for learning are but stationary. Should it be his good fortune to forget nothing, still he remains at the point of educational advancement where he was […] when he quit the scalpel.

The anatomist on the other hand, continually hunched over cadavers, advances in his research, realizes his conjectures; each day brings something new, for in anatomy we do not yet know everything; and like practical medicine, it cannot be learned from books.

Notes

An Anatomist in the Century of Enlightenment

1. Richelot, vol. 1, 1839, p. 197.
2. Buffon, vol. 4, p. 5.
3. *Encyclopédie*, vol. 6, 1756, article about strangles (a contagious bacterial infection in horses), p. 74.
4. The *Encyclopédie* in 1765 included the revolutionary terms *comparative medicine* and *comparative surgery*. "Medicine is a branch of natural history, which also includes anatomy. There will never be a good theory of medicine without a thorough understanding of natural history, because one cannot understand the *economie* of man without knowing the conformation of the different animals; and in medicine, in which the most rapid progress has been made up to now, a comparative medicine and surgery are needed just as is a comparative anatomy." *Encyclopédie*, vol. 8, 1765, article "Natural History," p. 226.

Honoré Fragonard (1732–1799)

1. The first five paragraphs of this chapter are adapted from the chapter entitled "Grasse" (pp. 11–16) in Verly's thesis. The life of Honoré Fragonard has been extremely well studied by Verly, who in 1963 submitted his thesis for a doctorate in veterinary medicine titled "Honoré Fragonard, anatomiste, premier directeur de l'École d'Alfort." He was aided in his research by Yvonne Poulle-Drieux, author of remarkable publications on Fragonard in 1962. All references to the details of Fragonard's life are taken from this thesis.
2. Archives of Grasse, HH 10: "Réceptions des chirurgiens de 1742 à 1791," in Verly, 1963, pp. 15–16.
3. Dronne, 1965, p. 65.
4. Figure taken from the National Archives of France, "a receipt of payment to Sir Fragonard (December 30, 1763), Director of the School, for his appointment of eight months, at 1200 livres a year, totaling 800 livres." It was the highest salary paid to an employee of the school; Philibert Chabert, who joined the school in October of the same year as "garçon de forge," was paid only 468 livres.
5. This quotation, attributed to Arloing (1889), is reproduced in numerous documents. We have not found proof of its authenticity.
6. On this subject, Verly has found in the National Archives in Paris two very detailed documents (Arch. Nat., F^{10}, dossier 1254 à 1257, "comptabilité de l'école d'Alfort [1767 à 1774]"). It is primarily an account for the month of March 1766: "for the carriage from Lyon to here and payment to customs for the receipt of the skeleton of a goat; a human subject: 21 L; spirit of wine for the foetus of a horse and the placenta in two large jars; a human subject: 3 L; a human subject: 3 L 4 s; one pint of spirit of wine for the mould; four human subjects at 21 L each; for a monkey; for four pairs of enamel eyes: 3 L; for carriage of specimens to Charenton."
7. Chabert, Flandrin, Huzard, 1809 (1st ed. 1782), p. 23.
8. Bourgelat, 1777, article 14, part 2, p. 132.
9. Chabert. Paris, published by the widow Vallat-la-Chapelle, 1782. This work was published without its author's name, but we know that Chabert was the chief editor. It was reedited for the first time in 1790 under the names of Chabert, Flandrin, and Huzard, then underwent several new editions under the title "d'Instructions et observations sur les maladies des animaux domestiques" (Instructions and Observations on the Illnesses of the Domestic Animals).
10. Thiéry, 1785, p.129.
11. Rumpelt, 1802 (extract from *Bulletin de la Société Centrale de médecine vétérinaire*, 1905, p. 413).
12. Sander, 1783.
13. Extract from an account of Rudolphi's journey, *Bemerkungen aus dem Gebiet*, published in Berlin in 1804–1805.
14. Vène, 2001, p. 9.
15. MacGregor, 2007, p. 164.
16. It was thus that this establishment received in 1796 the beautiful collection of "insectes de Barbarie" from Pierre-François Guyot Desfontaines (1685–1745), the collection of African birds from François Levaillant (1753–1824) in 1797, the collection of Guyanese birds from Brocheton in 1798, and, the same year, the immense collection brought from the Antilles by Nicolas Thomas Baudin (1754–1803).
17. Taken from Verly, 1963, p. 38.
18. Archives of the Veterinary School of Alfort, article 107. Deposited in the Archives départementales du Val-de-Marne, Créteil.
19. Biographical information in the first four paragraphs of this section are from Verly, 1963, pp. 37–44.
20. Verly has located the case reports in the National Archives in Paris (Arch. Nat, T. 1 130).
21. Extract from Thouret's eulogy of Fragonard, April 5, 1799.
22. Le Breton, 1993.
23. Cited in Mandressi, 2003.
24. Their signatures are to be found side by side on nearly all the reports of

the temporary commission of the arts up until April 1795. One regularly specifies near his name his occupation of painter, while the other mentions only once (March 10, 1794) his profession of anatomist. However, Honoré Fragonard showed himself to be much more assiduous in the sessions of the commission than his cousin, who permanently gave up helping him after April–May 1795.

25. Guillaume, 1894, vol. 2, p. 327.
26. Deloche and Leniaud, 1989, p. 174.
27. Guillaume, 1894, vol. 2, p. 327.
28. Arch. nat., F17 1164, file 6, no. 3: "…I had found it [the cabinet] in good order and had observed only that the windows and doors did not shut and the blinds were in a bad state. Conducted into the dissection room, I saw that the ceiling was ready to give way and that they had been forced to shore it up at both ends to prevent that accident and to be not privy to the amphitheater, which is situated above: in that room I have found the pathologies of bone and pathologies of soft tissue preserved in spirit of wine, the forges in order, and the instruments kept with the greatest care, each piece is numbered, classified, described, and that description is entrusted to the students." In Poulle-Drieux, 1962, vol. 15, document 2, p. 155.
29. Arch. Nat., F 17 1238, file 1, no. 9, May 24, 1794. Contemporary title on the jacket: "Report of citizens Fragonard and Vicqdazir on the cabinet of anatomy at the *École nationale vétérinaire d'Alfort près de Charenton*." Cf. p. 147.
30. It was Mme. Yvonne Poulle-Drieux who discovered this inventory in the National Archives in Paris, where it was filed under the serial number F10 1294.
31. Guillaume, vol. 5, 1904, pp. 564–567.
32. An order of February 22, 1796, raised this salary to 9,000 livres, plus lodgings.
33. Corlieu, 1878.
34. Thouret, 1798, pp. 1–20.
35. Arch. Nat. F17 2281, cited by Michel Lemire, *Artistes et mortels* (Artists and Mortals) (Paris: Chabaud, 1990), 181.

36. The collection of the surgeon Desault, well-known anatomist and surgeon.
37. Prévost, 1901, p. 135 : "Account of the death of Fragonard. 19 *Germinal* year VII."
38. Corlieu, 1896, chapter 7, pp. 105–108, "I. Le chef des travaux anatomiques (the chief of anatomical works)."

From the Royal Veterinary College to Modern-Day Plastination

1. Harvey published his discovery in 1628 in his *Exercitatio anatomica de motu cordis et sanguinis in animalibus*.
2. This discovery was published in 1627, after Aselli's death, in *De lactibus sive lacteis venis*.
3. Vène, 2001, p. 16.
4. Molière, *The Imaginary Invalid*, act 2 scene 5.
5. Ruini, 1598.
6. Vallat, 1973, p. 16.
7. Héroard, 1599.
8. Foisil, 1989.
9. Bourgelat, 1750–1753.
10. Plate in volume 1, containing the knowledge of the horse considered from the exterior, and a short treatise theoretical and practical on shoeing.
11. Buffon, 1753, vol. 4, pp. 256–57. "I shall not speak of other ailments of the horse that I shall cover in *Natural History*, which gives an account of the maladies of the animal; however, I cannot complete the story of the horse without noting with some regret that the health of this most useful and precious animal has up until now been left to the care and the practice, often blind, of those without knowledge and without education. The 'Medicine' that the ancients have called 'Veterinary Medicine' is hardly known except as a name: I am persuaded that if some doctors turned their gaze from their own subject and made that study their principal object, they would soon be compensated by ample success; that not only would reward them, but rather than degrading themselves it would bring them much renown, and this Medicine should not be as conjectural and as difficult as the other: the nourishment, the morals,

the influence of sentiment, in a word being simpler in the animal than in Man, the ailments should also be less complicated, and as a consequence easier to diagnose and treat with success; without counting the complete freedom that one would have to learn by experience, to try new remedies, and to be able to arrive without fear and without reproach at an extensive knowledge in this area, in which one can even by analogy draw some useful inferences for the art of treating humans."
12. Stubbs, 1766.
13. Morrison, 2003.
14. Chabert, Flandrin, Huzard, 1803–1804, p. 7.
15. Chabert, Flandrin, Huzard, 1801–1802, p. 363.
16. Huzard, notice. Bourgelat, *Art vétérinaire ou médecine des animaux* (Veterinary Arts, or The Medicine of the Animal), 3rd edition. (Paris: Vallat-La-Chapelle, 1794).
17. Jean Girard. *Traité d'anatomie vétérinaire ou histoire abrégée de l'anatomie et de la physiologie des principaux animaux domestiques* (Treatise on Veterinary Anatomy, or The Abridged History of Anatomy and Physiology of the Main Domestic Animals). vol. 1, 2nd edition (Paris: Huzard, 1819), 24. This mention of Fragonard does not occur in the very similar phrase that appears in the first edition of 1807.
18. Arch. Nat., F 10 1294.
19. Monro, 1789, p. 679.
20. The paper by Alexander Monro describing the technique of injection served as the basis of the article "injecter" in the *Encyclopédie* (vol. 8, p. 746, December 1765), reprinted in the *Encyclopédie méthodique*. The paper in question had been translated into French and inserted into *Essais et observations de Médecine de la société d'Edinburgh* (Essays and Observations on Medicine in the Society of Edinburgh), vol. 1, article 9, p. 113 *et seq.*
21. Sue, 1748.
22. Tarin, 1750. This anonymous work is also sometimes attributed to Sue.
23. Sue, 1765. This edition was in the

collection of the Veterinary School of Alfort.

24. Rudolphi, 1805, p. 33.
25. Le Breton, 1993.
26. Rumpelt, 1802.
27. Sander, 1783.
28. Antoine-Joseph Dézallier d'Argenville, *L'Histoire naturelle éclaircie dans deux de ses parties principales, la lithologie et la conchyliologie.* (Natural History Explained in Two of Its Principal Parts: Lithology and Conchology) (Paris : De Bure aîné, 1742), 205–208.
29. Cited by Verly, 1963, p. 65. Extract from Dézallier d'Argenville, *La Conchyliologie ou Histoire naturelle des coquilles de mer, d'eau douce, terrestres et fossiles, avec un traité de la zoomorphose, ou représentation des animaux qui les habitent…* (*Conchyliologie*, or The Natural History of the Shells of the Sea, of Fresh Water, Terrestrial and Fossil, with a Treatise on the *Zoomorphose*, or Representations of the Animals that Inhabit Them…) (Paris: G. de Bure, 1780), 223–24.
30. Anon., 1863.
31. Cited by Verly, 1963, p. 63:
   "cabinet in room 8:
   no. 266: Donkey (*Equus asinus Linné*), dried foot. Fragmented specimen originating from the Veterinary School of Alfort
   no. 267: Donkey, dried foot. Given by Veterinary School of Alfort and made by Fragonard
   no. 270: Cow, dehydrated anterior foot. Prepared by Fragonard and given by the Veterinary School of Alfort
   no. 271: Cow, dehydrated posterior foot. Specimen prepared by Fragonard and given by the Veterinary School of Alfort."
32. Until 1991 the *écorchés* were kept in the first glass cabinet of the museum, which today contains the skulls. This placement gave them direct southern exposure to the sun. Glass fragments found under the clavicle of the *Man with a Mandible* show that at some point the glass ceiling of that cabinet collapsed.
33. These preparations are listed in the 1903 inventory preserved at the museum: "one myology and splanch-nology of a man and a horse: the horseman anatomized; myology and splanchnology of a man; injected and dehydrated preparation of a man standing upright, holding in the right hand an inferior maxilla of a horse; myology and splanchnology of a human arm; myology and splanchnology of a human leg; two injected preparations of human heads; four preparations of human foetuses; vascular system injected (probably human); myology of a monkey; myology and splanchnology of a llama; myology and splanchnology of an antelope; two dried preparations of porpoises; umbilical cord and foetal membranes of a mare; preparation of the anterior part of a sheep with four horns."
34. Vicq d'Azyr, 1793–1794, p. 37.

The Écorchés

1. Rudolphi, 1805.
2. Ibid., p. 33.
2. Ibid.
4. Ibid., "…the reins, like the whip that the rider held, have since disappeared."
5. Inventory of 1794, no. 1447.
6. Ibid.
7. Ibid., no. 1430 and no. 1431.
8. For an example, see "Honoré Fragonard, l'écorché vif " ("Honoré Fragonard, Flayed Alive"), *Newlook* (March 1989): 26–31. Page 29 is headlined "For 'the amazon,' he dug up the body of a young girl."
9. Sander, 1783.
10. The Bible, Book of Judges 15:14–17. "And when he came to Lehi, the Philistines shouted against him: and the Spirit of the Lord came mightily on him, and the cords that were on his arms became as flax that was burnt with fire, and his bands loosed from off his hands. And he found a new jawbone of an ass, and put forth his hand, and took it, and slew a thousand men therewith.
   "And Samson said, With the jawbone of an ass, heaps on heaps, with the jaw of an ass have I slain a thousand men.
   "And it came to pass, when he had made an end of speaking, that he cast away the jawbone out of his hand, and called that place Ramathlehi."
11. This cannot be the bust of the young girl listed under no. 1160, since the latter shows the arms and the heart.
12. In the National Archives, item 215 from File 1379.
13. Buffon. "The nilgai," supplement to vol. 6, 1782, pp. 101–115.
14. Buffon, "The llama and the alpaca," vol. 13, 1765, pp. 16–33.
15. Buffon, "The llama," supplement to vol. 6, 1782, pp. 204–207.
16. Louis Petit de Bachaumont, 1784, p. 18.

Yesterday and Today: From *Écorchés* to Plastinates

1. For example dioramas, which were invented by Louis-Jacques-Mandé Daguerre in the 1820s, were among the common means of exhibition in which theatricality was as much part of the presentation of curiosities of nature as was showing prowess in new display techniques.
2. With today's scanners and ultrasound, it is easy to overlook the genuine revolution in both theory and therapeutic approach brought by the discovery of X-rays in 1895 by the German physicist Roentgen. Allowing access to the interior body of a living individual, the X-ray bypassed reliance on dissection of the dead, and in fact marked the beginning of the gradual decline of the cadaver as the sole source of observation.
3. Among recent reviews and analyses see Rafaël Mandressi, 2003; Le Breton, 1993.
4. Several million people worldwide have already visited this type of exhibition.
5. On this point see Philippe Ariès, "L'anatomie pour tous" ("Anatomy for All"), *La mort ensauvagée*, vol. 2 of *L'homme devant la mort* (Facing Death) (Paris: Seuil, 1985), 74.
6. Gunther von Hagens describes his exhibitions as "edutainment."
7. After immersion in successive baths of acetone to remove water and fat, the specimen is impregnated under a vacuum with synthetic liquid resins, the most common being silicone, which then hardens on contact with a catalytic vapor. While the high quality of the specimens

obtained had never been achieved in the past, the technique itself is based on exactly the same principles of impregnation that were practiced beforehand with paraffin wax and later, from the 1950s on, with polyethylene glycol.

8. The status of the anatomist has evolved in line with that of the discipline. The anatomists of the eighteenth century were generally surgeons, as was John Hunter, or doctors, such as his obstetrician brother William. It was not until the nineteenth century that the great European universities entrusted the training of their medical students to teachers who specialized in anatomy and dedicated themselves to the teaching profession rather than the medical profession. Today, anatomy is often considered a secondary subject and taught according to systems by the relevant specialists in medicine or surgery.

9. Article 3 of the law of November 15, 1887, on the freedom of funerals.

10. Article 16-1 of the Civil Code.

11. As such, in France, the collections held by establishments designated *musée de France*, such as the Musée de l'Ecole vétérinaire d'Alfort, benefit from the principles of inalienability, indefeasibility, and exemption under a law passed January 4, 2002.

Preparation Techniques for Dry Anatomical Specimens

1. This point is fundamental, for if the pigment formed lumps it would have no chance of passing through the small vessels. It was for that reason that Sue (1765, p. 25) advised against using cinnabar (mercuric sulphide, the common ore of mercury), the naturally occurring form of vermilion, which precipitates once injected.

2. Sue, 1765, chapter 1, section 6, p. 31.

3. Before the adoption of the metric system in April 1795, the reference unit in France was the Paris pound, which equated to 489.5 g. The pound was divided into 16 ounces of 8 drams; each dram was 72 grains.

4. One ounce was equivalent to 30.6 g.

5. Sue, 1765, chapter 1, section 5, p. 29.

6. Ibid.

7. Ibid., section 11, pp. 45–46.

8. Ibid., section 14, p. 49.

9. Ibid., section 15, pp. 50–51.

10. Ibid., section 18, p. 55.

11. Ibid., section 20, pp. 59–61.

12. Ibid., section 16, pp. 52–53.

13. Ibid.

14. From the Latin *menstruum*, meaning monthly. A solvent that took approximately one month to work, at moderate heat.

15. "Spirit of turpentine": turpentine essence. In old chemistry, the term "spirit" indicates a product of distillation.

16. John Hunter, British surgeon and anatomist.

17. "Turpentine of Venice": turpentine formerly extracted in the Alps from the Larch, *Larix europœa*.

18. Aqueous solution of highly concentrated hydrochloric acid.

19. Nitric acid.

20. Sue, 1765, chapter 6, section 1, "The preparation of the pericranium, the dura mater, and the dural sinuses," pp. 222–28.

21. Ibid., chapter 10, p. 256.

22. Felix Vicq d'Azyr (1793, p. 37) emphasizes his many interests.

23. In Ernest-Alexandre Lauth, *Nouveau manuel de l'anatomiste* (New Manual of the Anatomist) (Paris: Levrault, 1835), 751.

24. Acetic acid.

25. Mercury chloride [$HgCl_2$].

26. The exact nature of this varnish is not known. Evidently Sue did not refer to an exact product, only the color and the solvent. Pierre Boitard (1853) mentioned a white varnish used for dry anatomical preparations. This was composed of Canada balsam (90 g), spirit of turpentine (90 g), and mastic varnish (60 g) (p. 422). Canada balsam, or Canada turpentine, is the transparent and colorless resin from the balsam fir (*Abies balsamea*). Mastic varnish, according to him, also known as yellow varnish (p. 423), is formed from powdered mastic (the resin of the mastic tree *Pistacia lentiscus*, a tree common on the south shore of the Mediterranean) dissolved in essence of turpentine. Lauth (1835) recommended using either a varnish

of spirit alcohol, oil of roasted nuts with lead oxide, or, his favorite, copal resin varnish.

27. Composed of copper carbonate.

28. Composed of iron cyanide.

29. "White lead": lead carbonate also known as "ceruse" or "spirits of Saturn."

30. Extract of the resin of the maritime pine (*Pinus pinaster*), a transparent, colorless liquid.

31. Monro, 1789, p. 677. "Spirit of wine": *eau-de-vie*, or spirit alcohol.

32. Mercuric sulphate.

33. The presence of an essential oil is attested by the discovery of camphor and eucalyptol in the injections. As Pierre Boitard stipulated in his treatise (p. 425), fatty and resinous substances can be "dissolved in alcohol, the fats, the wax." The essence of turpentine was the most commonly used, but its unpleasant odor made him prefer the essence of lemon or lavender (oil of aspic). Fragonard, born into a family of perfumiers, would have known these products well.

34. Monro, 1789, p. 681.

35. Gannal, 1838, p. 218. "Tallow, 5 parts; Burgundy resin (spruce), 2 parts; oil of olives or nuts, 2 parts, liquid turpentine and coloring matter dissolved in essential oil, 1 part."

36. Boitard, 1853, p. 427. Boitard proposed the following mixture for injecting the great vessels: tallow (153 g), Burgundy resin (60 g), oil of olives or nuts (60 g), liquid turpentine loaded with coloring matter (30 g).

   Pierre Boitard was an erudite naturalist who published a great many works on various topics in natural history.

37. Larousse, 1866–1877, vol. 9 (H–K), p. 700

   "For the arteries, a mixture of tallow and white pitch [white Burgundy pitch] is used, to which is added, after it has been melted and passed through a cloth, essence of turpentine into which is dissolved some lamp black."

Bibliography

Theses and Dissertations

BOURLAUD, Marie-Anne. "Les Techniques anatomiques de base et leurs modalités. Essai d'élaboration d'un index bibliographique en vue de l'enseignement et de la recherche." DVM thesis, Veterinary School of Alfort, 1971, no. 22.

DRONNE, Michel. "Bertin et l'élevage français au xviiie siècle." DVM thesis, Veterinary School of Alfort, 1965, no.49.

VALLAT, François. "Histoire de l'illustration dans l'anatomie vétérinaire française." DVM thesis, Veterinary School of Lyon, 1973.

VERLY, Pierre-Louis. "Honoré Fragonard, anatomiste, premier directeur de l'École d'Alfort." DVM thesis, Veterinary School of Alfort, 1963, no. 30.

Articles and Books before 1950

ANON. *Livret du musée d'anatomie normal de la faculté de médecine de Paris.* Paris: Masson et fils, 1863.

ARLOING, Saturnin. *Le Berceau de l'enseignement vétérinaire.* Lyon: imprimerie Pitrat aîné, 1889.

BACHAUMONT, Louis Petit de. *Mémoires secrets pour servir à l'histoire de la République des lettres en France, depuis 1752 jusqu'à nos jours ; ou Journal d'un observateur.* London: John Adamson, 1784, vol. 17.

BÉGUILLET, Edme. *Histoire de Paris, avec la description de ses plus beaux monuments, dessinés et gravés en taille-douce.* Paris: Duchesne, 1780, 3 vols.

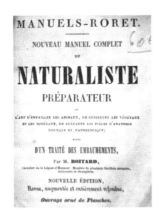

BOITARD, Pierre. *Nouveau manuel complet du naturaliste préparateur.* Paris: Librairie encyclopédique de Roret, 1853.

BOULEY, Henri. "Éloge de M. Jean Girard," read in the public meeting held at the School of Alfort, October 12, 1854, Paris.

BOURGELAT, Claude. *Élémens d'hippiatrique ou nouveaux principes sur la connoissance et sur la médecine des chevaux. Tome premier, contenant la connaissance du cheval considéré extérieurement, et un traité abbrégé théorique et pratique sur la ferrure.* Lyon: Henri Declaustre et les Frères Duplain, 1750.

———. *Élémens d'hippiatrique ou nouveaux principes sur la connaissance et sur la médecine des chevaux. Tome second, première partie, contenant un abbrégé hyposteologique, myologique et angéiologique.* Lyon: Henri Declaustre et les Frères Duplain, 1751. *Tome second, seconde partie, contenant un précis anatomique de la tête et de la poitrine du Cheval.* Lyon: Henri Declaustre et les Frères Duplain, 1753.

———. *Règlement pour les Écoles royales vétérinaires de France.* Paris: Imprimerie royale, 1777.

BREDIN, Louis. *Observations en réponse au mémoire de M. Lafosse sur l'École royale vétérinaire d'Alfort,* 1790.

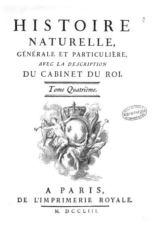

BUFFON, Georges Leclerc, (comte de). "Le lama et le paco." *Histoire naturelle générale et particulière avec la description du cabinet du Roi*, vol. 13. Paris: Imprimerie royale, 1765, pp. 16–33.

———. "Le lama." *Histoire naturelle générale et particulière avec la description du cabinet du Roi, supplément au tome VI*. Paris: Imprimerie royale, 1782, pp. 204–207.

CHABERT, Philibert. *Almanach vétérinaire, contenant l'histoire abrégée des progrès de la médecine des animaux depuis l'établissement des écoles vétérinaires en France*. Paris: chez la veuve Vallat-la-Chapelle, 1782.

———, Pierre FLANDRIN, and Jean-Baptiste HUZARD. *Instructions et observations sur les maladies des animaux domestiques*, vol. 1, 1782, 4th edition. Paris: Mme Huzard, 1809.

———. *Instructions et observations sur les maladies des animaux domestiques*, vol. 3, 1792, 2nd edition. Paris: Mme Huzard, an VII (1799).

———. *Instructions et observations sur les maladies des animaux domestiques*, vol. 4, 1793, 2nd edition. Paris: Mme Huzard, an X (1801–1802).

———. *Instructions et observations sur les maladies des animaux domestiques*, vol. 5, 1793, 2nd edition. Paris: Mme Huzard, an XII (1803–1804).

CORLIEU, Auguste. *Le Chef des travaux anatomiques de la Faculté de médecine de Paris*. Paris: Imprimerie nationale, 1878.

———. *Centenaire de la faculté de Médecine de Paris (1794–1894)*. Paris: Imprimerie nationale, 1896.

DELEUZE, Joseph Philippe François. *Histoire et description du Muséum royal d'histoire naturelle*. Paris: chez M. A. Royer, au Jardin du roi, 1823.

FREDERICQ, Léon. "Sur quelques procédés nouveaux de préparation des pièces anatomiques sèches." *Bulletin de l'Académie royale de Belgique*, 1876, p. 41.

———. "Conservation à sec des tissus mous par la paraffine." *Gazette médicale de France*, 1879, series 6, p. 45.

GANNAL, Jacques Nicolas. *Histoire des embaumements et de la préparation des pièges [sic] anatomie normale, d'anatomie pathologique et d'histoire naturelle ; suivie de procédés nouveaux*. Paris: chez Ferra, 1838.

GIRARD, Jean. *Traité d'anatomie des animaux domestiques*. Paris: chez Huzard, 1807, 2 vols.

GOUBAUX, Armand. "Fragonard." *Bulletin de la Société centrale de médecine vétérinaire*, 1888.

GUILLAUME, James. *Procès-verbaux du Comité d'instruction publique de la Convention Nationale*, vol. 5, 1904, pp. 564–67.

HÉROARD, Jean. *Hippostologie, c'est-à-dire "Discours des os du cheval."* Paris: chez Mamert Patisson imprimeur ordinaire du Roy, 1599.

LAFOSSE, Étienne Philippe. *Cours d'hippiatrique ou Traité complet de la Médecine des Chevaux*. Paris: chez Edme, 1772.

LAROUSSE, Pierre. *Grand Dictionnaire universel du xixe siècle : français, historique, géographique, mythologique, bibliographique*. Paris: Administration du grand Dictionnaire universel, 1866–1877, vol. 9 (H–K), p. 700.

MONRO, Alexander. "Art des préparations anatomiques." *Encyclopédie méthodique, médecine*, t. VI. Paris: chez Panckouck, 1789, p. 679.

MOULÉ, Léon. "Correspondance de Claude Bourgelat, fondateur des écoles vétérinaires." *Bulletin de la Société centrale de médecine vétérinaire*, 1911, p. 343, pp. 390–91.

———. "Correspondance de Claude Bourgelat, fondateur des écoles vétérinaires." *Bulletin de la Société centrale de médecine vétérinaire*, 1912, pp. 63–64.

———. "Correspondance de Claude Bourgelat, fondateur des écoles vétérinaires." *Bulletin de la Société centrale de médecine vétérinaire*, 1916, pp. 286–88.

NEUMANN, Louis-Georges. *Biographies vétérinaires*. Paris: chez Asselin et Houzeau, 1896.

PRÉVOST, A. *L'École de Santé de Paris (1794–1809)*. Bibliothèque historique de la France médicale, 1901.

RAILLIET, Alcide, and Léon MOULÉ. *Histoire de l'École d'Alfort*. Paris: chez Asselin et Houzeau, 1908.

RICHELOT, Gustave-Antoine. *Œuvres complètes de John Hunter*. Paris: Labé et Didot, 1839.

RUDOLPHI, Karl Asmund. *Bemerkungen aus dem Gebiet der Naturgeschichte, Medicin und Thierarzneykunde, auf einer Reise durch einen Theil von Deutschland, Holland und Frankreich*, part 2. Berlin: Realschulbuchhandlung, 1805.

RUINI, Carlo. *Anatomia del Cavallo, infermita e suoi remedii, etc.*, 1st edition. Bologne: Eredi Rossi, 1598.

RUMPELT, Georg Ludwig. *Veterinarische und ökonomische Mittheilungen von einer Reise durch einige Provinzen Deutschlands, Hollands, Englands, Frankreichs und der Schweiz.* Dresden: Walther, 1802.

SANDER, Heinrich. *Beschreibung seiner Reisen durch Frankreich, die Niederlande, Holland, Deutschland und Italien; in Beziehung auf Menschenkenntnis, Industrie, Litteratur und Naturkunde insonderheit.* Leipzig: Jacobäer, 1783.

STUBBS, George. *The Anatomy of the Horse, Including a Particular Description of the Bones, Cartilages, Muscles, Fascias, Ligaments, Nerves, Arteries, Veins and Glands.* London: printed by J. Purser for the author, 1766.

SUE, Jean-Joseph. *Abrégé de l'anatomie du corps de l'homme*, par M. Sue, chirurgien et professeur adjoint en anatomie de l'Académie royale de peinture et sculpture. Paris: chez Simon, 1748.

———. *Anthropotomie ou l'art de disséquer, d'embaumer et de conserver les parties du corps humain, & c*, 2nd edition revised and greatly expanded. Paris: chez l'auteur & chez Cavellier, 1765. This edition was among the original holdings of the School of Alfort.

TARIN, Pierre. *Anthropotomie ou l'art de disséquer, d'embaumer et de conserver les parties du corps humain, & c.* Paris: chez Briasson, 1750.

THIÉRY, Luc Vincent. *Almanach du voyageur à Paris, contenant une description sommaire mais exacte, de tous les Monumens, Chefs-d'œuvre des Arts, Établissemens utiles, & autres objets de curiosité que renferme cette Capitale : Ouvrage utile aux Citoyens, & indispensable pour l'Étranger.* Paris: chez Hardouin et Gattey, 1785.

THOURET, Michel Augustin. *De l'état actuel de l'École de Santé de Paris.* Paris: chez Didot Jeune, an VI (1798).

VICQ D'AZYR, Félix. *Instruction sur la manière d'inventorier et de conserver, dans toute l'étendue de la République, tous les objets qui peuvent servir aux arts, aux sciences, et à l'enseignement, proposé par la Commission temporaire des arts, et adopté par le Comité d'instruction publique de la Convention nationale.* Paris, an II (1793–1794), p. 37.

Articles and Works after 1950

BRION, Marcel. *Art fantastique.* Paris: Albin Michel, 1961.

CADOT, Laure. *En chair et en os : le cadavre au musée.* Thesis research, École du Louvre, 2006, p. 99.

DELOCHE, Bernard, and Jean-Michel LENIAUD. "La culture des sans-culottes." *Le Premier Dossier du patrimoine 1789–1798.* Paris and Montpellier: Les éditions de Paris, Les Presses du Languedoc, 1989, p. 174.

ELLENBERGER, Michel. *L'Autre Fragonard.* Paris: Jupilles, 1981.

FOISIL, Madeleine. *Journal de Jean Héroard, médecin de Louis XIII.* Paris: Fayard, 1989, 2 vols.

LE BRETON, David. *La Chair à vif. Usages médicaux et mondains du corps humain.* Paris: Éditions Métailié, 1993.

MACGREGOR, Arthur. *Curiosity and Enlightenment.* New Haven: Yale University Press, 2007.

MAMMERICK, Marc. *Claude Bourgelat, avocat des vétérinaires.* Bruxelles, self-published, 1971.

MANDRESSI, Rafaël. *Le Regard de l'anatomiste. Dissections et invention du corps en Occident.* Paris: Le Seuil, 2003.

MORRISON, Venetia. *George Stubbs, l'essentiel de son œuvre.* Paris: Succès du Livre, 2003.

POULLE-DRIEUX, Yvonne. "Honoré Fragonard et le cabinet d'anatomie de l'École d'Alfort." *Revue d'histoire des sciences*, vol. 15, 1962, pp. 141–62.

ROSSIGNOL, Lorraine. "Amour à mort sous silicone." *Le Monde*, April 17, 2009.

VÈNE, Magalie. *Écorchés. L'exploration du corps. XIVe–XVIIIe siècle.* Paris: Albin Michel/BnF, 2001.

ACKNOWLEDGMENTS

My deepest gratitude is to Pierre-Louis Verly, whose thesis "Honoré Fragonard, Anatomist, First Director of the School of Alfort," has been invaluable to this book and to the museum.

I am grateful to Mrs. Poulle-Drieux, who released Fragonard from the shadows of history. Her work has nourished the development of this book.

Deep gratitude also to Professors Jean-Paul Mialot, Bernard Toma, André-Laurent Parodi, Robert Moraillon, and Jean-Pierre Cotard, respectively director and former directors of the National Veterinary School of Alfort, for their continued support of the Fragonard Museum and its curator. Thanks, too, to Alain Hénault, the memory of the museum.

For images, we have benefited from the generous support of Valérie Brousselle and Éric Jingeaux, the archives at Val-de-Marne; Guy Cobolet, Jean-François Vincent, Stéphanie Charreaux and Estelle Lambert, Bibliothèque Inter-universitaire of Paris; Philippe Comar, Department of Morphology, École nationale supérieure des beaux-arts, Paris; Professor Marjanne E. Everts and Lisanne van der Voort, Veterinary Faculty, University of Utrecht; Luc Gomel, curator at the University Montpellier 2, Montpellier; Elisabeth Grison, Virginie Willaime, and Anne-Christine Romary, librarians, National Veterinary School of Alfort; Sigrid Kohlmann, librarian, University of Erlangen-Nuremberg, Germany; Delilah Walle and Bart Grob, Museum Boerhaave, Leiden, Netherlands.

We also thank the photographers Patrick Forget, Patrick Landmann, Guillaume de Laubier, Jean-Marc Simmonet, and Pierre-Emmanuel Weck, and also Gilles Sacksick, Robert Henry, Henry Chateau, and Georges-Claude Boniteau, as well as the Institute for Plastination, Heidelberg, Germany.

For the handsome English-language edition of this book, we are grateful to publishers Laura Lindgren and Ken Swezey of Blast Books, and especially to Laura for her skillful layout and her careful, thorough editing of the translation for which we are grateful to Philip Adds writing on a tight deadline. Thanks, too, to Joanna Ebenstein for recommending Blast Books to us.

Deepest gratitude to all who have given us their help and advice.

PHOTOGRAPHY CREDITS

Gilles Sacksick, 2010, p. 12; Patrick Forget/Sagaphoto, pp. 9, 18, 51, 64, 66, 67, 90, 91, 92–93, 95, 96–97, 98, 101, 103, 106, 107, 109, 111, 114, 116, 118–121, 136–137, 138, 154; Patrick Landmann, pp. 88–89, 99, 104, 105, 112–113, 117; Guillaume de Laubier, p. 41; Jean-Marc Simonet, pp. 26, 28; Pierre-Emmanuel Weck, pp. 10, 84–85, 94, 95; Créteil, Archives départementales du Val-de-Marne, p. 24; Archives de l'école vétérinaire d'Alfort conservées aux Archives départementales du Val-de-Marne scan: Eric Jingeaux, p. 38; Erlangen, Sigrid Kohlmann, Handschriftenabteilung/ Graphische Sammlung, Universitätsbibliothek, p. 31; Heidelberg, Gunther von Hagens's BODY WORLDS, Institute for Plastination, pp. 128, 134; La Haye, Musée royal des Peintures Mauritshuis, pp. 16–17; Leiden, Museum Boerhaave, p. 30 ; Maisons-Alfort, ENVA, bibliothèque/C. Degueurce, pp. 11, 25, 32, 33, 34, 35, 43, 60–63, 69, 70, 108, 110, 139, 140, 155–157; ENVA, musée Fragonard/C. Degueurce, pp. 3, 6, 42, 47, 73, 87, 100, 102, 124, 141–146, 155-157; ENVA/C. Degueurce, p. 21; ENVA/H. Chateau & C. Degueurce, p. 83; Montpellier, Université Montpellier 2/C. Degueurce, p. 123, 141; Paris, Archives nationales, pp. 22, 46; Bibliothèque inter-universitaire de médecine, pp. 13, 54, 56, 57, 131; Ecole nationale supérieure des beaux-arts, p. 69; Institut médico-légal/C. Degueurce, p. 122; Musée Delmas-Orfila-Rouvière, université Paris V/C. Degueurce, pp. 76, 79; Muséum national d'histoire naturelle, p. 59; RMN/Daniel Arnaudet, p. 8; RMN/Château de Versailles/Gérard Blot, p. 21, RMN/Gérard Blot, p. 45; RMN/tous droits réservés, p. 73; RMN/agence Bulloz, p. 36; Rouen, musée Flaubert et d'histoire de la médecine/ CHU-Hôpitaux de Rouen, p. 59; Utrecht, Lisanne van der Voort, Multimedia Department, Faculty of Veterinary Medicine, University of Utrecht, pp. 80, 81.